Heat Transfer and Fluid Flow in Rotating Coolant Channels

W. David Morris, B.Sc.(Eng.), Ph.D., C.Eng., F.I.Mech.E., F.I.Prod.E.

Fenner Professor of Mechanical Engineering, The University of Hull, England

RESEARCH STUDIES PRESS
A DIVISION OF JOHN WILEY & SONS LTD
Chichester · New York · Brisbane · Toronto · Singapore

RESEARCH STUDIES PRESS

Editorial Office:
8 Willian Way, Letchworth, Herts SG6 2HG, England

Copyright © 1981, by John Wiley & Sons, Ltd.

All rights reserved.

No part of this book may be reproduced by any means, nor
transmitted, nor translated into a machine language
without the written permission of the publisher.

British Library Cataloguing in Publication Data:

Morris, W. David
  Heat transfer and fluid flow in rotating coolant
  channels.—(Mechanical engineering research studies;
  2)
  1. Turbomachines—Fluid dynamics
  2. Thermodynamics
  I. Title
  621.8'11   TJ267

  ISBN 0 471 10121 4

Printed in Great Britain

IN MEMORY OF MY FATHER

FOREWORD

This monograph reviews the information currently available in the technical press concerning the influence of rotation on flow and heat transfer in ducts which are constrained to rotate about a prescribed axis. Since this topic plays an important role in the overall performance of cooling systems designed for rotating components, notably prime movers, the information has been presented in a manner intended to satisfy the requirements of the potential industrial user as well as the academic researcher. In this respect design-type recommendations are made at all possible instances throughout the monograph.

Although the text draws heavily from the results of research with which I have personally been involved over the years I should like to acknowledge the stimulation which I have had from studying the results of other workers from all over the world and the contribution this has made to my own understanding of the subject.

The major part of my own contribution to this research area was undertaken while I was a member of faculty at the University of Sussex and I should like to record my appreciation to my friend and former colleague Professor F J Bayley, Director of the SRC Thermo-Fluid Mechanics Research Centre at the University for his constant support and encouragement.

I should also like to acknowledge the assistance given to me by Mrs S Langdale for her care and patience during the preparation of the manuscript and likewise to Miss P Cherry for help with the illustrations.

W. David Morris
University of Hull
1981.

TABLE OF CONTENTS

6   FLOW AND HEAT TRANSFER IN CIRCULAR-SECTIONED TUBES WHICH
      ROTATE ABOUT AN ORTHOGONAL AXIS WITH LAMINAR OR TURBULENT
      FLOW.

NOMENCLATURE

English Symbols

a	tube radius
$\underline{a}$	position vector
a_ϕ	coefficient
A	constant
A_1, A_2, A_3, A_4	nodal finite difference coefficients
b	side dimension of square-sectioned duct
B	constant
B_N, B_S, B_W, B_E	nodal finite difference coefficients
C	constant
C_f^*	excess friction factor
C_{fo}	Blasius friction factor for stationary tube
C_{fR}	Blasius friction factor for rotating tube
C_p	constant pressure specific heat
C_N, C_S, C_W, C_E	nodal finite difference coefficients
d	tube diameter
d_ϕ	coefficient
D	velocity term
D_p	nodal finite difference coefficient
E	function
E_1, E_2, E_3, E_4, E_5	coefficients
f	function
$\underline{f}$	acceleration vector
$\underline{f_o}$	acceleration vector
F	function
$\underline{F}$	body force vector
F_x, F_y, F_z	body force components
F_1, F_2, F_3, F_4, F_5	coefficients
G	function

Gz	Graetz number
H	eccentricity
$H_1,H_2,H_3,H_4,H_5,H_6,H_7$	coefficients
$\underline{i}$	unit vector
$\underline{j}$	unit vector
J_a	rotational Reynolds number (based on tube radius)
J_b	rotational Reynolds number (based on side dimension for a square-sectioned duct)
J_d	rotational Reynolds number (based on tube diameter
k	thermal conductivity/turbulent kinematic energy
$\underline{k}$	unit vector
L	length of tube
Nu	Nusselt number
Nu_m	mean value of Nusselt number
Nu_o	Nusselt number for stationary tube
Nu_z	local value of Nusselt number
$Nu_{\infty,b}$	bulk developed Nusselt number
$Nu_{\infty,u}$	unweighted developed Nusselt number
p	pressure
P'	pressure perturbation
Pr	Prandtl number
Pr_t	turbulent Prandtl number
$\dot{q}$	heat flux
$\dot{Q}$	heat transfer rate
r	radial coordinate
$\underline{r}$	position vector
$\underline{r_o}$	position vector
R	non-dimensional radial coordinate
Ra_b	Rayleigh number based on wall-bulk temperature difference

Ra_τ	Rayleigh number based on axial temperature gradient
Re	through flow Reynolds number
Re_c	transitional Reynolds number
Re_p	pseudo Reynolds number
Ro	Rossby number
Ro'	alternative form of Rossby number
R_o	particular nodal value of non-dimensional radial coordinate
R_W, R_E, R_P	nodal values of non-dimensional radial coordinate
t	time
T	temperature
T_b	bulk temperature
T_m	unweighted mean temperature
T_o	reference temperature
T_w	wall temperature
u	velocity component
U	non-dimensional velocity component
U_c	non-dimensional velocity component in core region
v	velocity component
$\underline{v}$	velocity vector
$\underline{v}_o$	velocity vector
V	non-dimensional velocity component
V_c	non-dimensional velocity component in core region
w	velocity component
w_m	mean axial velocity
W	non-dimensional velocity component
W_c	non-dimensional velocity component in core region
$W_{c\Delta}$	velocity component at edge of boundary layer

W_o	zero order non-dimensional velocity
W_1	first order non-dimensional velocity
W_2	second order non-dimensional velocity
x	Cartesian coordinate
X	non-dimensional Cartesian coordinate/ function
y	Cartesian coordinate
Y	non-dimensional Cartesian coordinate
z	Cartesian coordinate
Z	non-dimensional Cartesian coordinate

Greek Symbols

α	thermal diffusivity
β	volume expansion coefficient
γ	pressure gradient
Δ	non-dimensional boundary layer thickness (hydrodynamic)
Δ_T	non-dimensional boundary layer thickness (thermal)
δ	boundary layer thickness (hydrodynamic)
δ_T	boundary layer thickness (thermal)
ϵ	eccentricity parameter (H/d)
$\underline{\epsilon}$	rate of turbulent dissipation
ϵ_a	eccentricity parameter (H/a)
ϵ_b	eccentricity parameter (H/b)
η	non-dimensional temperature
η_c	non-dimensional temperature in core region
$\eta_{c\Delta}$	non-dimensional temperature at edge of boundary layer
η_o	particular nodal value of non-dimensional temperature
θ	angular coordinate
$\theta_N, \theta_S, \theta_P$	nodal values of angular coordinate

λ	non-dimensional pressure gradient
μ	viscosity
μ_e	effective viscosity
υ	kinematic viscosity
ξ	vorticity
$\underline{\xi}$	vorticity vector
ξ_{wall}	wall value of vorticity
ρ	density
ρ'	density perturbation
ρ_o	reference density
σ	normal stress
τ	temperature gradient/tangential shear stress
Γ	function
ϕ	scalar function
Φ	general dependent variable
$\Phi_N, \Phi_S, \Phi_W, \Phi_E, \Phi_P$	nodal values of generalised dependent variable
Ψ^*	stream function
Ψ_o	zero order stream function/particular nodal value
Ψ_1	first order stream function
Ψ_2	second order stream function
χ	pressure-type term
χ^*	non-dimensional pressure type term
χ_c^*	non-dimensional pressure type term in the core region
$\underline{\omega}$	angular velocity vector
Ω	angular velocity

CHAPTER 1

THE STRATEGIC AIM OF THE MONOGRAPH

1.1 Introduction

The ability to predict the performance of an engineering product
with good accuracy at an early stage in the design process has
obvious commercial advantages. This is to ensure, in the first
instance, that new concepts are adequately assessed for feasibility
and later for the reduction of expensive development programmes. The
identification of a potentially successful design depends typically
on a systematic application of technical analysis, manufacturing
knowledge and organisational skills. Further, successful design
depends not only on the basic concept being sound, but also on the
reliable performance of certain vital components, or sub-assemblies
necessary to support the original concept.

As the demand for more technically sophisticated products has
increased, for example due to advances in aerospace requirements,
many areas where knowledge is sparse or indeed non-existent have
been highlighted. This in turn has stimulated increased activity in
fundamental and applied engineering research strategically directed
at eventual product improvement. The attendant "information explo-
sion" which has occurred in the technical literature often makes it
difficult for the industrial user to have easy and concise access to
the available information that his specific needs demand.

It may be justifiably argued therefore that there are certain important research topics which should be periodically reviewed and
reported in a manner suitable for direct industrial usage. With this
objective in mind the present monograph reviews, critically appraises
and makes design recommendations on one selected topic which has been
the subject of active research in recent years. Specifically the
topic reviewed deals with the assessment of heat transfer and flow
resistance in cooling channels designed for rotating components. Before embarking on the details of the influence of rotation on cooling
system performance two important practical applications where this
research material has direct applicability will be described in order
to justify the importance and industrial relevance of the monograph.

1.2 Cooling Problems in the Rotors of Electrical Machines

Significant economic advantages may be realised if electrical power is produced using turbine-driven generators which have a relatively high output per machine typically in excess of 500 MW. This is due to the fact that the capital cost per unit power delivered is reduced if the total required plant output is generated with a notionally small number of highly rated turbine-generator sets. Further, under these circumstances the overall plant efficiency tends to be higher.

As well as the electrical principles upon which the performance of these highly loaded machines depend, reliable operation over a commercially acceptable lifespan can only be achieved if considerable attention is given to mechanical aspects of the overall design envisaged. In this respect careful attention to considerations of stress, rotor vibrations and thermal behaviour becomes vital.

The rotor speed is dependent upon the frequency at which power is to be generated and this consequently limits the diameter of the rotor to a value compatible with the mechanical strength properties of the materials of construction used. Also the overall length of the rotor is restricted by the need to avoid excessive vibrational problems and the need to be able to machine and generally manipulate the rotor forgings used. Once the physical size of the machine has been limited by the mechanical constraints mentioned above the only remaining way by which the power output may be increased is via increases in the electrical and magnetic loadings in the stator and rotor of the machine. The implied consequential increase in the absolute level of general inefficiencies within the machine necessitates a reliable cooling system to be incorporated into the fundamental design concept. This is to ensure that the thermal losses are dissipated at temperature levels compatible with an acceptable lifespan of the electrical insulation materials used in present day generator construction.

Details of any cooling system design depend not only on the total thermal power to be dissipated but also on the relative distribution of this loss between the stator and rotor. Although opinions vary concerning the future size trend of single unit turbogenerators (see for example Hawley (1969)*, Arnold (1970), Creek (1977)), figure 1.1 illustrates the typical tendencies over the last three decades.

With generators of relatively large output it is customary to cool both stator and rotor windings by circulating a suitable coolant through channels located inside the windings themselves. This technique, which utilizes hollow conductors, is referred to as direct cooling and hydrogen, due to its attractive thermodynamic and transport properties, has been successfully used to advantage.

As power outputs have increased, say notionally in excess of 500 MW there has been a move towards fully water cooled machines by some manufacturers. Stator cooling using water is now normal for machines with outputs in excess of 200 MW because design, development and operational experience has brought about the attendant confidence in reliability. However only fairly recently have the mechanical problems associated with maintaining an internal water flow in the rotor

* References cited will be listed alphabetically at the end of the monograph.

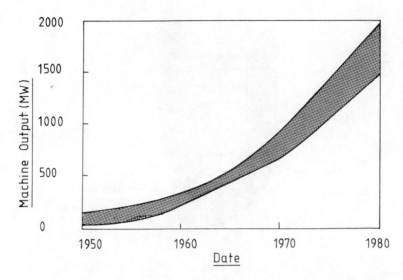

FIG. 1.1   THE TREND FOR SINGLE-MACHINE TURBOGENERATOR
OUTPUTS OVER THE YEARS.

been solved with sufficient reliability to warrant the thermal advantages of using water-cooled rotors, see for example Bennett (1968),
and Pohl (1973).

Cooling system layouts for electrical rotors will vary in points of
detail in accordance with the required design loadings and the con-
ceptual preferences of various manufacturers.   It is evident nevertheless, that certain flow features will be fundamental and common to
all.   Thus, for example, the windings of the rotor, with their internal cooling channels, are mounted in slots machined axially along
the rotor forging.   In practice, a number of mutually insulated copper windings are located in each slot and held in position with
appropriate packing and wedges as shown schematically in figure 1.2.
This figure also illustrates the point that in practice axial cooling
holes having a variety of cross sectional shapes are commonly used.

The sequences of schematic flow circuits shown in figure 1.3 demonstrate the variety of options commonly available to the designer to
distribute the coolant to the rotor windings.   In figure 1.3a the
slot windings utilize a predominantly axial flow along the copper
windings, a situation which sometimes arises with water cooled rotors.
Figure 1.3b depicts an alternative system which again utilizes a
mainly axial through flow of coolant.   This flow system differs in
that relatively short axial cooling paths are used with each individual path supplied with coolant from a distribution channel located
beneath the copper region and usually referred to as a subslot.   The
heated coolant from each of these relatively short axial channels is
subsequently exhausted into the so called "air gap" between the rotor

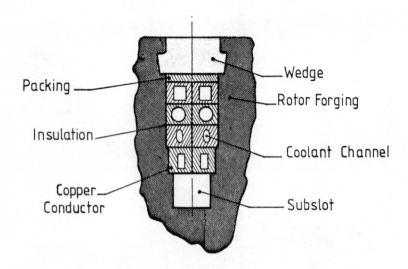

FIG. 1.2   TYPICAL LOCATION OF COPPER WINDINGS IN
            ROTOR SLOT SHOWING POSSIBLE INTERNAL
            COOLING HOLE GEOMETRIES.

periphery and the inner-surface of the stator. Figure 1.3c shows a
coolant distribution system used to good effect on machines of fairly
low power ratings. The coolant is again distributed to the copper
via a subslot region but in this instance is discharged into the air-
gap via radial holes punched in the windings.

The choice of the assorted flow circuit geometries is closely related to the machine rating and choice of coolant made. However, it
should be noted that any coolant circuit configuration selected for
the rotor windings must also include suitable delivery and exhausting
arrangements for the coolant. This typically involves a combination
of bends, expansions, contractions, plenum chambers, etc. Reliable
prediction of the thermal and hydrodynamic characteristics of the
resulting complex flow circuits must essentially take into account
the manner in which rotation of the circuit influences these characteristics. This feature may be quantified by noting that a 10 K increase in the temperature level of the insulation separating rotor
conductors, based on a nominal 373 K operating temperature, will
effectively half the operational life of the insulation.

1.3 Cooling Problems in the Rotor Blades of High Temperature
     Turbines.

It is well known that there are economic advantages which may be
realised by operating steam or gas turbines at high maximum cycle
temperatures. Upper limits on temperature are imposed by the mechanical properties of the materials used in construction and in this

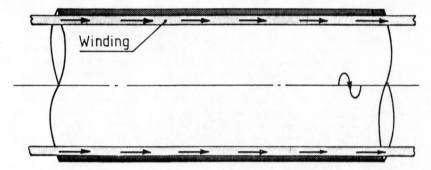

1.3a   Windings with axial cooling channel.

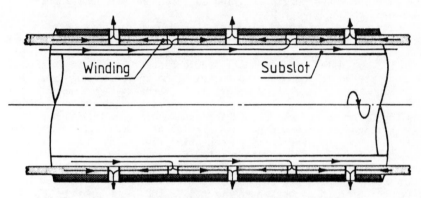

1.3b   Windings with short axial cooling channels.

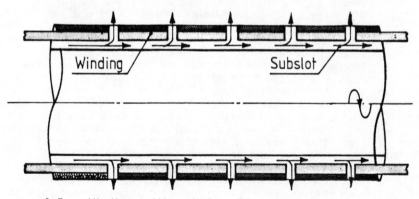

1.3c   Windings with radial cooling channels.

FIG. 1.3   TYPICAL DIRECT COOLING SYSTEMS FOR
TURBOGENERATOR ROTOR WINDINGS.

respect the turbine rotor blades are particularly critical. Typically rotor blades must be sufficiently strong to sustain the loads resulting from a centripetal acceleration field of the order of 10^4 g whilst simultaneously exposed to a thermal environment hot enough to cause the blade material to glow red. Further, blades must be capable of resisting fatigue, thermal shock, creep and oxidation.

Figure 1.4 shows the typical improvement in turbine entry temperature resulting from progressive development over the years for aero-type applications. Prior to the early 1960s there was a progressive increase in turbine entry temperature of about 10 K per year. This trend has trebled since about 1960 due to the widespread adoption of air-cooling for turbine blades. Air cooling has obvious advantages of convenience and availability and figure 1.5 shows a commonly used commercial arrangement. Here both nozzle and rotor blades are cooled by the use of hollow blades through which the cooling air flows. In the case shown the nozzle trailing edge regions are additionally film cooled as shown.

A wide variety of rotor blade cooling geometries have been in successful commercial use and figure 1.6 shows the commonly adopted and relatively simple series of spanwise cooling passages involved with the so-called conventionally cooled rotor blade. In this case coolant is fed to the blade root region and thence through radial holes machined in the blade material. Alternative arrangements include film cooled augmentation of the trailing edges, transpiration cooling using porous blades, impingement cooling and various multi-pass arrangements. However, irrespective of the detailed nature of the flow geometry used it is an essential feature of the system that the coolant is constrained to rotate with the blade itself.

For a given blade design and a specified operational duty the distribution of temperature in the blade material may be predicted by solving the heat conduction equation over the blade profile of interest, and hence an estimate of operational life made.

It is necessary, however to prescribe the thermal boundary conditions which prevail over the external surface of the blade and also at the internal surfaces of any coolant channels in order to complete the solution. The accuracy of the prediction is crucially dependent on the accuracy of these prescribed thermal boundary conditions and in this respect considerable attention has justifiably been directed at the nature of the flow and heat transfer characteristics of the gas flow over the external surface of the blade. It has been customary to assume that the thermal boundary condition at the internal coolant passages may be adequately specified using correlations for laminar or turbulent flow, as appropriate, which were originally derived for equivalent stationary flow geometries. It is interesting in this respect to note that Fox (1974) has demonstrated that a 10% reduction in heat transfer at the internal surface of a conventionally cooled turbine rotor blade can typically result in an increase of blade mean temperature which produces an attendant sevenfold decrease in creep life together with a twofold reduction in corrosion life. It would appear therefore that a design procedure which does not include the effect of rotation on flow and heat transfer in any rotor blade coolant passages used could give suspect life predictions.

7

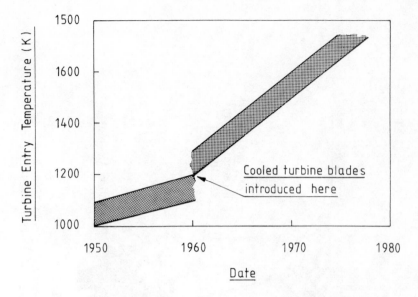

FIG. 1.4   THE TREND FOR TURBINE ENTRY TEMPERATURE
INCREASES OVER THE YEARS FOR AERO-TYPE
APPLICATIONS.

1.4 The Strategic Aim

The last two sections of this chapter have demonstrated a number of
instances in practice where the effect of rotation on the hydrodynamic
and thermal characteristics of channel-type flows may have important
consequences on the performance of cooling systems designed for prime
movers.   In view of its practical importance it may be argued that
there is a serious need to critically review the assorted literature
available on this generalised topic and present it in a form which
bridges the gap between the academic researcher on the one hand and
the eventual industrial user on the other.   It is with this aim in
mind that the present monograph has been written.

8

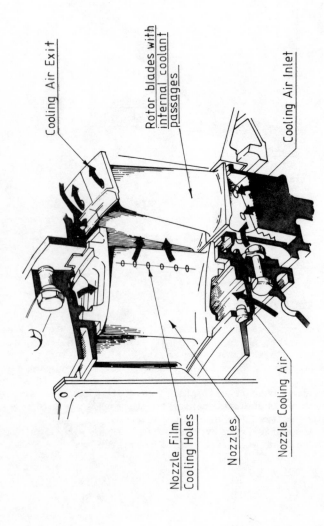

Cooling Air Exit

Rotor blades with internal coolant passages

Cooling Air Inlet

Nozzle Film Cooling Holes

Nozzles

Nozzle Cooling Air

FIG. 1.5   TYPICAL COOLING ARRANGEMENT FOR GAS TURBINE BLADES (with acknowledgement to Rolls Royce (1971) Ltd).

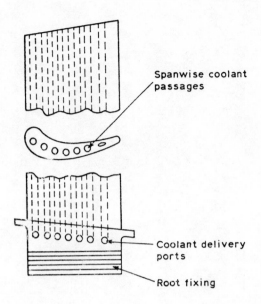

Spanwise coolant passages

Coolant delivery ports

Root fixing

FIG. 1.6   TYPICAL ARRANGEMENT OF SPANWISE COOLING
PASSAGES IN A GAS TURBINE ROTOR BLADE.

CHAPTER 2

REVIEW OF FUNDAMENTAL PRINCIPLES AND THEIR

RELATIONSHIP TO FLOW IN ROTATING DUCTS

2.1 Introduction

The last chapter described some important practical situations
where fluid flows in rotating ducts and made the point that an assessment of how the rotation affects the flow field, and the associated
convective process in the case of heated flows, is necessary for
reliable performance predictions to be made. We now address our-
selves in general terms to a discussion of the mechanisms by which
duct rotation influences the flow in relation to the non-rotating
case. Subsequent to the treatment of the general fundamental principles a detailed study of specific flow geometries will be presented
in later chapters.

Motion is governed by Newton's three laws of motion. The so-called
second law relates the resultant force acting on a body to the change
in momentum experienced by the body. However for the second law to
be valid it is necessary to ensure that the motion concerned is referred to an inertial frame of reference or, in other words, a reference system which permits uniform motion of the body to be specified when there is no resultant force acting. This requires a reference frame which is either at rest in space or moving with uniform
velocity.

If the motion of a fluid particle moving inside a stationary duct
is required, a convenient reference frame may be postulated which is
fixed in relation to the duct itself, for example a simple Cartesian
coordinate system may be used. Because the duct is stationary such a
reference frame will be inertial and the customary application of
Newton's second law of motion may be made. However, if the duct is
rotating and/or translating a reference frame fixed to the duct is no
longer inertial. Under these circumstances correction terms must be
applied to the usual mathematical embodiment of Newton's second law
in order to preserve its validity. Although in principle an inertial
frame of reference can always be defined it is often more convenient
to adopt a non-inertial frame which moves with the duct. The correction terms required to account for the non-inertial nature of a reference frame attached to a rotating duct will now be presented and a
number of examples, important for later chapters, given.

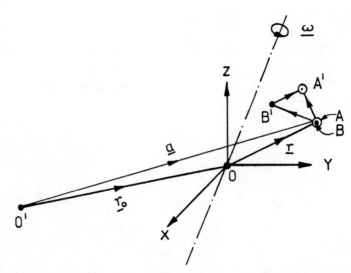

FIG. 2.1   TYPICAL TRANSLATING AND ROTATING FRAME OF
REFERENCE.

2.2   Motion Referred to Non-inertial Reference Frames

Consider a particle, A, moving in space. At some instant in time,
t, the particle has a position vector, $\underline{a}$, with reference to an inertial coordinate frame which has its origin at O', see figure 2.1. A
rectangular Cartesian frame of reference, OXYZ, has its origin at O
and the position vector of O with reference to the inertial reference
O' is $\underline{r}_o$. Suppose the frame OXYZ rotates about O, as indicated in

figure 2.1, with an angular velocity vector $\underline{\omega}$. In the general case,
when $\underline{r}_o$ is not constant, the reference frame is simultaneously trans-

lating and rotating. Let the particle have position vector, $\underline{r}$, with
reference to the frame OXYZ. After the elapse of an interval of time,
δt, the particle A has moved to A' in space and during this period
the point B, originally coincident with A but fixed relative to the
rotating/translating frame, has moved to B' as shown in the figure.
Suppose for the present that the vector $\underline{r}_o$ is constant. From
figure 2.1 we observe that

$$\underline{AA}' = \underline{BB}' + \underline{B'A}' \tag{2.1}$$

where, over the time interval t to $(t + \delta t)$, $\underline{AA}'$ represents the
change in position of the particle in space, $\underline{B'A}'$ represents the
change in the position of the particle relative to the rotating frame
and $\underline{BB}'$ represents the change in position of the point fixed relative
to the rotating frame and coincident with the particle at time t.

The path $\underline{BB}'$ may be expressed as the vector product of the angular displacement, $\underline{\omega}\delta t$, of the point B and the position vector $\underline{r}$ to give

$$\underline{BB}' = (\underline{\omega} \wedge \underline{r})\delta t \qquad (2.2)$$

Thus the velocity vector, $\underline{v}$, of the particle A in space may be determined using equations (2.1) and (2.2) as

$$\underline{v} = \lim_{\delta t \to o}\left[\frac{\underline{AA'}}{\delta t}\right] = \lim_{\delta t \to o}\left[\frac{\underline{B'A'}}{\delta t}\right] + (\underline{\omega} \wedge \underline{r}) \qquad (2.3)$$

The term $\lim_{\delta t \to o}\left[\dfrac{\underline{AA'}}{\delta t}\right]$ expresses the rate of change of $\underline{r}$ in absolute

terms and any fixed frame of reference at O may be used to locate the

particle concerned. It is convenient to adopt the notation $\dfrac{d}{dt}$ to

refer to time-wise differentiation with respect to any fixed reference frame having its origin at O. In a similar manner $\lim_{\delta t \to o}\left[\dfrac{\underline{B'A'}}{\delta t}\right]$

represents the rate of change of $\underline{r}$ relative to the rotating frame of

reference with its origin at O. Let the notation $\dfrac{\partial}{\partial t}$ be used to

signify this relative time-wise differentiation. Equation (2.3) may now be conveniently written as

$$\underline{v} = \frac{d\underline{r}}{dt} = \frac{\partial \underline{r}}{\partial t} + (\underline{\omega} \wedge \underline{r}) \qquad (2.4)$$

There is no reason to prevent the above argument being extended to vectors other than $\underline{r}$ so that equation (2.4) may be treated as an operator such that

$$\frac{d}{dt} = \frac{\partial}{\partial t} + \underline{\omega} \wedge \qquad (2.5)$$

Note carefully that the operators $\dfrac{d}{dt}$ and $\dfrac{\partial}{\partial t}$ must be consistently used in the context described above.

The acceleration, $\underline{f}$, in space of the particle A may be derived easily using the identity specified by equation (2.5). Thus

$$\underline{f} = \frac{d\underline{v}}{dt} = \frac{\partial \underline{v}}{\partial t} + (\underline{\omega} \wedge \underline{v}) \qquad (2.6)$$

which, on expansion using equation (2.4), gives

$$\underline{f} = \frac{\partial^2 \underline{r}}{\partial t^2} + 2\left[\underline{\omega} \wedge \frac{\partial \underline{r}}{\partial t}\right] + \left[\frac{\partial \underline{\omega}}{\partial t} \wedge \underline{r}\right] + \underline{\omega} \wedge (\underline{\omega} \wedge \underline{r}) \qquad (2.7)$$

Equations (2.4) and (2.7), respectively representing the absolute velocity and acceleration in space of the particle A, were derived on the assumption that O was at rest. For the more general case where the origin O translates with velocity v_0 and acceleration $\underline{f}_0$ with respect to the inertial frame at O' then these quantities must simply be added to equations (2.4) and (2.7) respectively to give the general result

$$\underline{v} = \frac{\partial \underline{r}}{\partial t} + (\underline{\omega} \wedge \underline{r}) + \underline{v}_0 \qquad (2.8)$$

and

$$\underline{f} = \frac{\partial^2 \underline{r}}{\partial t^2} + 2\left[\underline{\omega} \wedge \frac{\partial \underline{r}}{\partial t}\right] + \left[\frac{\partial \underline{\omega}}{\partial t} \wedge \underline{r}\right] + \left[\underline{\omega} \wedge (\underline{\omega} \wedge \underline{r})\right] + \underline{f}_0 \qquad (2.9)$$

We note from equation (2.9) that three correction terms involving the angular velocity of the rotating frame must be used to determine the acceleration with reference to the inertial frame OXYZ. Also a fourth correction term involving the translational acceleration of O must be incorporated if the origin of the reference frame is not itself fixed.

The term $\underline{\omega} \wedge (\underline{\omega} \wedge \underline{r})$ is known as the CENTRIPETAL ACCELERATION*

whereas the term $2\left[\underline{\omega} \wedge \frac{\partial \underline{r}}{\partial t}\right]$ is known as the CORIOLIS ACCELERATION**.

These two terms occur whenever motion is referred to a reference frame which is itself rotating. If the reference frame is rotating with non-uniform angular velocity then an additional term $\left[\frac{\partial \underline{\omega}}{\partial t} \wedge \underline{r}\right]$

is required to account for the angular acceleration of the frame.

By way of exemplification equation (2.9) will be evaluated for two special cases which will be encountered in later chapters.

* derived from Latin and implying "centre fleeing".
** the acceleration vector given by equation (2.9) was originally derived by G.G. Coriolis (1829).

2.3 Acceleration of a Fluid Particle Moving in a Straight Duct which Rotates About an Axis Parallel to the Main Flow Direction.

Case 1 : Motion referred to a Cartesian frame fixed relative
to the duct.

Consider the rotating flow geometry depicted in figure 2.2 which
consists of a straight duct of arbitrary cross section which is constrained to rotate with uniform angular velocity, Ω, about an axis
parallel to the duct itself. This will be referred to as parallel-
mode rotation. Let the motion of a typical particle of fluid located
at A be referred to the rectangular Cartesian frame OXYZ shown. This
frame rotates with the duct and the eccentricity between the Z-axis
(the direction of flow) and the rotational axis is H.
The position vector, $\underline{r}$, of A relative to the origin O is

$$\underline{r} = x\underline{i} + y\underline{j} + z\underline{k} \tag{2.10}$$

where $\underline{i}$, $\underline{j}$ and $\underline{k}$ are unit vectors in the OX, OY and OZ directions
respectively and x, y and z are the coordinates of A with respect to
OXYZ. Also, the angular velocity vector of the frame OXYZ in space
is

$$\underline{\omega} = \Omega\underline{k} \tag{2.11}$$

The translational acceleration, $\underline{f}_o$, of O relative to an inertial
frame located at O' is

$$\underline{f}_o = -\Omega^2 H\underline{i} \tag{2.12}$$

Because the angular velocity of the duct is constant in this case,
substitution of equations (2.10), (2.11) and (2.12) into equation
(2.9) yields

$$\underline{f} = \frac{\partial^2}{\partial t^2} [x\underline{i} + y\underline{j} + z\underline{k}] + 2 [\Omega\underline{k} \wedge \frac{\partial}{\partial t} (x\underline{i} + y\underline{j} + z\underline{k})]$$

$$+ [\Omega\underline{k} \wedge (\Omega\underline{k} \wedge (x\underline{i} + y\underline{j} + z\underline{k}))] - \Omega^2 H\underline{i} \tag{2.13}$$

Expansion of equation (2.13) yields after some manipulative algebra

$$\underline{f} = [\frac{\partial^2 x}{\partial t^2} - 2\Omega\frac{\partial y}{\partial t} - \Omega^2(H + x)] \underline{i}$$

$$+ [\frac{\partial^2 y}{\partial t^2} + 2\Omega\frac{\partial x}{\partial t} - \Omega^2 y] \underline{j}$$

$$+ \frac{\partial^2 z}{\partial t^2} \underline{k} \tag{2.14}$$

16

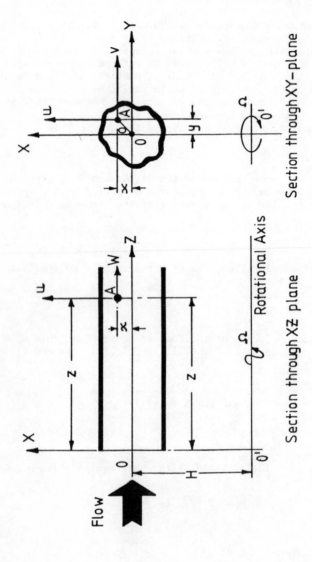

FIG. 2.2 PARALLEL-MODE ROTATION REFERRED TO A CARTESIAN FRAME OF REFERENCE

In deriving equation (2.14) the following point of detail is important. When evaluating terms involving $\frac{\partial}{\partial t}$ the unit vectors $\underline{i}$, $\underline{j}$ and $\underline{k}$ are treated as constants by virtue of the special usage of this time-wise derivative as described earlier. If u, v and w are the velocity components of A in the directions OX, OY and OZ respectively then equation (2.14) becomes

$$\underline{f} = [\frac{\partial u}{\partial t} - 2\Omega v - \Omega^2(H + x)] \; \underline{i}$$

$$+ [\frac{\partial v}{\partial t} + 2\Omega u - \Omega^2 y] \; \underline{j} \; + \frac{\partial w}{\partial t} \; \underline{k} \tag{2.15}$$

Equation (2.15) gives the correction terms required to specify the acceleration vector for a fluid particle flowing in a duct constrained to rotate in the manner shown in figure 2.2. Note that the correction terms arise in this case from the Coriolis effect, the centripetal effect and the implied translational acceleration of the origin used.

Case 2 : Motion referred to a cylindrical polar frame fixed
              relative to the duct.

Suppose now that the previous example is re-examined but with the motion referred to a cylindrical polar coordinate system which is assumed fixed relative to the duct. In this instance the z-direction is still along the duct itself but the cross stream location of the typical particle at A is located by reference to r and θ as shown in figure 2.3. Let $\underline{i}$, $\underline{j}$ and $\underline{k}$ represent unit vectors in the r, θ and z directions with the positive sense of $\underline{j}$ being taken along the tangent in the increasing θ-direction.

With respect to the origin O, the position vector, $\underline{r}$ of the point A is

$$\underline{r} = r\underline{i} + z\underline{k} \tag{2.16}$$

The angular velocity vector, $\underline{\omega}$, of the reference frame is

$$\underline{\omega} = (\Omega + \dot{\theta}) \; \underline{k} \tag{2.17}$$

where $\dot{\theta}$ implies the time rate of change of the angle θ. Note the important feature, that although the tube is rotating with uniform angular velocity there is an implied angular acceleration of the reference frame since θ is changing as the position vector tracks the particle. The translational acceleration of the origin O is

$$\underline{f_o} = - \Omega^2 H \cos \theta \; \underline{i} \; + \; \Omega^2 H \sin \theta \; \underline{j} \tag{2.18}$$

Equations (2.16), (2.17) and (2.18) may be substituted, as before, into equation (2.9) to give

18

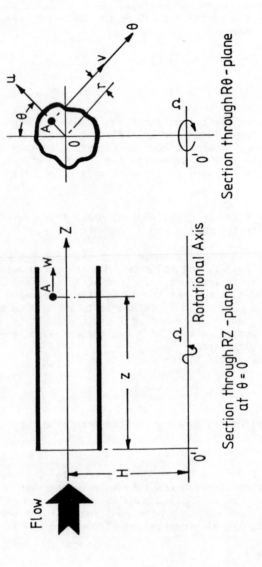

Section through RZ-plane
at θ = 0

Section through Rθ-plane

FIG. 2.3   PARALLEL-MODE ROTATION REFERRED TO A POLAR FRAME OF REFERENCE

$$\underline{f} = \frac{\partial^2}{\partial t^2} [r\underline{i} + z\underline{k}] + 2 [(\Omega+\dot\theta) \underline{k} \wedge \frac{\partial}{\partial t} (r\underline{i} + z\underline{k})]$$

$$+ [\frac{\partial(\Omega+\dot\theta)k}{\partial t} \wedge (r\underline{i} + z\underline{k})]$$

$$+ [(\Omega+\dot\theta)\underline{k} \wedge ((\Omega+\dot\theta)\underline{k} \wedge (r\underline{i} + z\underline{k}))]$$

$$- \Omega^2 H \cos\theta \, \underline{i} + \Omega^2 H \sin\theta \, \underline{j} \tag{2.19}$$

Expansion of (2.19) gives

$$\underline{f} = \left[\left(\frac{\partial^2 r}{\partial t^2} - r\dot\theta^2 \right) - 2\Omega r\dot\theta - \Omega^2(r + H\cos\theta) \right] \underline{i}$$

$$+ \left[\left(r\ddot\theta + 2\frac{\partial r}{\partial t}\dot\theta \right) + 2\Omega\frac{\partial r}{\partial t} + \Omega^2 H \sin\theta \right] \underline{j}$$

$$+ \frac{\partial^2 z}{\partial t^2} \underline{k} \tag{2.20}$$

If the velocity components in the r, θ and z directions are u, v and w respectively, where $v = r\dot\theta$, equation (2.20) becomes

$$\underline{f} = \left[(\frac{\partial u}{\partial t} - \frac{v^2}{r}) - 2\Omega v - \Omega^2 (r + H\cos\theta) \right] \underline{i}$$

$$+ \left[(\frac{\partial v}{\partial t} + \frac{uv}{r}) + 2\Omega u + \Omega^2 H \sin\theta \right] \underline{j}$$

$$+ \frac{\partial w}{\partial t} \underline{k} \tag{2.21}$$

The following points of detail are worthy of note. The terms $-\frac{v^2}{r}$ and $\frac{uv}{r}$ arise from the fact that the use of polar cylindrical coordinates has an implicit angular velocity component $\dot\theta\underline{k}$ and these terms exist even when the tube itself is stationary. Further, these terms will always appear when an r, θ, z frame is used. The acceleration component $-2\Omega v$ in the radial direction has the structure of a Coriolis term but in actual fact has its origin in the centripetal term. Equation (2.21) must be used in the application of Newton's Second Law if the motion is referred to a cylindrical polar frame attached to the tube.

2.4 Acceleration of a Fluid Particle Moving in a Straight Duct Which Rotates About an Axis Perpendicular to the Main Flow Direction.

Case 1 : Motion referred to a Cartesian frame fixed relative to the duct.

Figure 2.4 illustrates the flow geometry considered. A straight duct of arbitrary cross section is constrained to rotate with uniform angular velocity, Ω, about an axis which is perpendicular to the duct. This will be referred to as orthogonal-mode rotation. Fluid flows through the duct in the direction moving outwards from the axis of rotation as shown. Let the motion of a typical particle A be referred to the Cartesian frame OXYZ where the origin O is located at a distance, H, from the axis of rotation. Again the frame of reference OXYZ rotates with the duct. If the coordinates of A with reference to the specified frame are x, y and z and the unit vectors in these directions are $\underline{i}$, $\underline{j}$ and $\underline{k}$ respectively, then the position vector, $\underline{r}$, of A relative to O is

$$\underline{r} = x\underline{i} + y\underline{j} + z\underline{k} \tag{2.22}$$

Further, in this case, the angular velocity vector $\underline{\omega}$, of the reference frame is

$$\underline{\omega} = \Omega\underline{j} \tag{2.23}$$

and the translational acceleration, $\underline{f}_o$, of the origin O is

$$\underline{f}_o = -\Omega^2 H \underline{k} \tag{2.24}$$

The acceleration of the particle, A, now follows by substituting equations (2.22), (2.23) and (2.24) into equation (2.9) and noting that the velocity components of A relative to the tube are u, v and w in the x, y and z direction respectively. Thus

$$\underline{f} = (\frac{\partial u}{\partial t} + 2\Omega w - \Omega^2 x) \underline{i}$$

$$+ \frac{\partial v}{\partial t} \underline{j}$$

$$+ (\frac{\partial w}{\partial t} - \Omega^2(H + z)) \underline{k} \tag{2.25}$$

Case 2 : Motion referred to a cylindrical polar frame fixed relative to the duct.

As a final example, the motion of the particle A within the orthogonally rotating duct is referred to the polar frame of reference r, θ and z shown in figure 2.5. For this case, the position vector, $\underline{r}$, of A, the angular velocity vector, $\underline{\omega}$, of the rotating frame and the translational acceleration vector, $\underline{f}_o$, of the origin are given

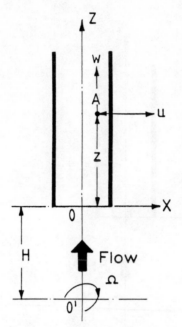

Section through XZ-plane

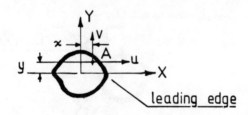

Section through XY-plane

FIG. 2.4   ORTHOGONAL-MODE ROTATION REFERRED TO A
CARTESIAN FRAME OF REFERENCE.

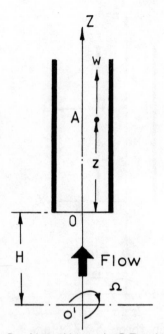

Section through RZ-plane

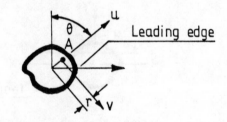

Section through Rθ-plane

FIG. 2.5   ORTHOGONAL-MODE ROTATION REFERRED TO A
POLAR FRAME OF REFERENCE.

by

$$\underline{r} = r\underline{i} + z\underline{k} \qquad (2.26)$$

$$\underline{\omega} = \Omega \cos \theta \underline{i} - \Omega \sin \theta \underline{j} + \dot{\theta}\underline{k} \qquad (2.27)$$

$$\underline{f}_o = - \Omega^2 H \underline{k} \qquad (2.28)$$

where $\underline{i}$, $\underline{j}$ and $\underline{k}$ are the unit vectors in the usual directions and H is the eccentricity of the selected origin.

Evaluation of equation (2.9) for this geometry gives the appropriate acceleration vector necessary for application in Newton's Second Law of Motion as

$$
\begin{aligned}
\underline{f} =\ & \left[(\frac{\partial u}{\partial t} - \frac{v^2}{r}) - 2\Omega w \sin \theta - \Omega^2 r \sin^2 \theta \right] \underline{i} \\
& + \left[(\frac{\partial v}{\partial t} + \frac{uv}{r}) - 2\Omega w \cos \theta - \Omega^2 r \sin \theta \cos \theta \right] \underline{j} \\
& + \left[\frac{\partial w}{\partial t} + 2\Omega (v \cos \theta + u \sin \theta) - \Omega^2 (z + H) \right] \underline{k} \qquad (2.29)
\end{aligned}
$$

where, as before, u, v and w are the velocity components in the coordinate directions r, θ, and z.

A detailed examination of the algebraic manipulations leading to equation (2.29) permits the following observations. The terms involving the square of the angular velocity are centripetal except for that due to the translation acceleration of the origin. A simple transformation of the z-direction coordinate to the intersection of the axis of rotation and the extension of the z-axis identically eliminates the eccentricity H. In the radial and tangential directions the terms which are proportional to the angular velocity of the tube are Coriolis in origin. In the z-direction the term $2\Omega u \sin \theta$ is also Coriolis in origin but the term $2\Omega v \cos \theta$ arises from the combined influence of the centripetal acceleration term and the angular acceleration term. The use of a cylindrical polar frame, even with a non-rotating duct, again produces the terms $\frac{v^2}{r}$ and $\frac{uv}{r}$.

2.5 Duct Rotation and the Creation of Vorticity.

2.5.1 Opening Remarks

In the previous sections of this chapter it has been demonstrated how additional terms must be included in the determination of the acceleration of a particle moving with respect to a rotating and translating reference frame. The qualitative effect of these additional acceleration terms, with their implied force implications, on flow behaviour in a rotating duct will now be discussed. To simplify this qualitative discussion the case of steady laminar constant property flow will be treated. We begin with a brief review of the laminar momentum conservation laws for motion referred to an inertial frame. Note that specific situations dealing with turbulent flow will be presented in later chapters.

2.5.2 The Momentum Conservation Equations for Constant Property Flow

Consider an element of fluid referred to an inertial rectangular Cartesian frame XYZ as shown in figure 2.6. Due to viscous action a combination of normal and shear stresses act on the element which, together with any body forces acting, can sustain fluid motion in accordance with Newton's laws.

The surface and body forces acting at a point in the fluid are summarised in figure 2.6 and Table 2.1 indicates the nomenclature adopted. Note that the subscripting adopted with the stresses implies that the first subscript denotes the plane in which the shear acts whereas the second subscript indicates the direction of action. Normal stresses are deemed positive when operating in a tensile mode.

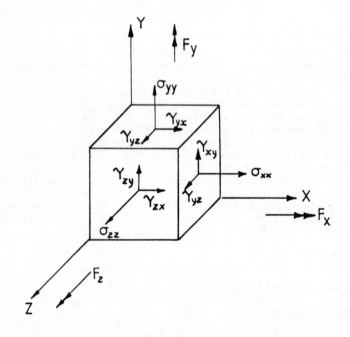

FIG. 2.6   STRESS AND BODY FORCE NOMENCLATURE

		Plane		Body Force
		X (yz)	Y (xz)	Z (xy)	
Coordinate Direction	X	σ_{xx}	τ_{yx}	τ_{zx}	F_x
	Y	τ_{xy}	σ_{yy}	τ_{zy}	F_y
	Z	τ_{xz}	τ_{yz}	σ_{zz}	F_z

σ = Normal shear stress component
τ = Tangential shear stress component
F = Body force component per unit mass of fluid

TABLE 2.1   TYPICAL FORCE DISTRIBUTION ON FLUID ELEMENT

The acceleration components for an inertial Cartesian reference frame are expressed via the usual "total derivative" of the velocity components which implies a differentiation following the motion of a specified particle.   Table 2.2 gives these acceleration components.

		Acceleration Component for Inertial Reference Frame						
Coordinate Direction	X	$\dfrac{Du}{Dt}$	$=$	$\dfrac{\partial u^*}{\partial t}$	$+$	$u\dfrac{\partial u}{\partial x}$	$+$	$v\dfrac{\partial u}{\partial y}$	$+$	$w\dfrac{\partial u}{\partial z}$
	Y	$\dfrac{Dv}{Dt}$	$=$	$\dfrac{\partial v}{\partial t}$	$+$	$u\dfrac{\partial v}{\partial x}$	$+$	$v\dfrac{\partial v}{\partial y}$	$+$	$w\dfrac{\partial v}{\partial z}$
	Z	$\dfrac{Dw}{Dt}$	$=$	$\dfrac{\partial w}{\partial t}$	$+$	$u\dfrac{\partial w}{\partial x}$	$+$	$v\dfrac{\partial w}{\partial y}$	$+$	$w\dfrac{\partial w}{\partial z}$

u, v and w refer to velocity components in the X, Y and Z directions respectively.

TABLE 2.2   ACCELERATION COMPONENTS FOR AN INERTIAL CARTESIAN REFERENCE FRAME.

When all the forces acting on a fluid element are assembled and incorporated into Newton's Second Law of Motion the resulting momentum conservation equations may be summarised in Table 2.3.

In principle the three components of Newton's conservation of momentum principle listed in Table 2.3 are equally applicable to laminar or turbulent flow.   This is due to the fact that, at this stage, no direct link has been made between the normal and tangential stresses, the fluid properties and the flow field itself.   For a so-called

* For the remainder of this book the operator $\dfrac{\partial}{\partial t}$ will imply the usual partial derivative with respect to time at a fixed point referred to the reference frame and not in the special context of Section 2.2.

		Momentum Conservation Equation			
Coordinate Direction	X	$\rho \dfrac{Du}{Dt}$	$= \rho F_x$	$+ \dfrac{\partial \sigma_{xx}}{\partial x} + \dfrac{\partial \tau_{yx}}{\partial y}$	$+ \dfrac{\partial \tau_{zx}}{\partial z}$
	Y	$\rho \dfrac{Dv}{Dt}$	$= \rho F_y$	$+ \dfrac{\partial \tau_{xy}}{\partial x} + \dfrac{\partial \sigma_{yy}}{\partial y}$	$+ \dfrac{\partial \tau_{zy}}{\partial z}$
	Z	$\rho \dfrac{Dw}{Dt}$	$= \rho F_z$	$+ \dfrac{\partial \tau_{xz}}{\partial x} + \dfrac{\partial \tau_{yz}}{\partial y}$	$+ \dfrac{\partial \sigma_{zz}}{\partial z}$

TABLE 2.3   MOMENTUM CONSERVATION EQUATIONS

Newtonian fluid, with its so-called laminar flow, it is assumed that the stress field is related in a linear fashion to the rate of strain. The details of the mathematical and algebraic details implied in this assumption will not be treated here. However the salient results are summarised* in Table 2.4.

Normal Stresses	$\sigma_{xx} = -p + 2\mu \dfrac{\partial u}{\partial x} - \dfrac{2}{3}\mu \ \mathrm{div}\ \underline{v}$	
	$\sigma_{yy} = -p + 2\mu \dfrac{\partial v}{\partial y} - \dfrac{2}{3}\mu \ \mathrm{div}\ \underline{v}$	
	$\sigma_{zz} = -p + 2\mu \dfrac{\partial w}{\partial z} - \dfrac{2}{3}\mu \ \mathrm{div}\ \underline{v}$	
Tangential Stresses	$\tau_{xy} = \tau_{yx} = \mu\left(\dfrac{\partial v}{\partial x} + \dfrac{\partial u}{\partial y}\right)$	
	$\tau_{yz} = \tau_{zy} = \mu\left(\dfrac{\partial w}{\partial y} + \dfrac{\partial v}{\partial z}\right)$	
	$\tau_{zx} = \tau_{xz} = \mu\left(\dfrac{\partial u}{\partial z} + \dfrac{\partial w}{\partial x}\right)$	

p = Fluid pressure
μ = Molecular viscosity of fluid
$\underline{v}$ = Velocity vector

TABLE 2.4   NORMAL AND TANGENTIAL STRESS RELATIONSHIPS FOR A NEWTONIAN FLUID.

The stress relationships outlined in Table 2.4 may be inserted into the momentum conservation equations given in Table 2.3. Instead of treating each coordinate direction separately it is convenient at this stage to combine the three equations into one vectorial expression of momentum conservation for a Newtonian fluid. Thus

* Readers are referred to Schlichting (1968), Lamb (1945), for more detailed derivations.

$$\rho \, \frac{Dv}{Dt} = - \nabla p + \rho \underline{F} + \mu \nabla^2 \underline{v} + \frac{\mu}{3} \nabla (\text{div } \underline{v}) \qquad (2.30)$$

where ρ is the density of the fluid and

$$\underline{F} = F_x \underline{i} + F_y \underline{j} + F_z \underline{k} \qquad (2.31)$$

with $\underline{i}$, $\underline{j}$ and $\underline{k}$ being the unit vectors in the coordinate directions x, y and z respectively. Also the operators ∇ and ∇^2 are defined as

$$\nabla = \frac{\underline{i}}{} \frac{\partial}{\partial x} + \frac{\underline{j}}{} \frac{\partial}{\partial y} + \frac{\underline{k}}{} \frac{\partial}{\partial z} \qquad (2.32)$$

$$\nabla^2 = \frac{\partial^2}{\partial x^2} + \frac{\partial^2}{\partial y^2} + \frac{\partial^2}{\partial z^2} \qquad (2.33)$$

For incompressible flow the conservation of mass principle necessitates that div $\underline{v} = 0$ so that for this special case equation (2.30) reduces to

$$\frac{Dv}{Dt} = - \frac{1}{\rho} \nabla p + \underline{F} + \nu \nabla^2 \underline{v} \qquad (2.34)$$

where $\nu = \frac{\mu}{\rho}$ is the kinematic viscosity of the fluid considered.

2.5.3 The Vorticity Equation for Unheated Flow in a Rotating Duct.

A convenient way to illustrate the manner in which rotation of a duct affects flow in that duct is to consider the creation of vorticity. Vorticity is a measure of the angular velocity which a particle of fluid has at a point in the flow. Suppose for example, that a small spherical element in the flow could be suddenly solidified. If the solid sphere is found to rotate then it is said to have vorticity. The vorticity is conventionally defined so that it is equal to double the implied angular velocity of the solid sphere considered. The formal mathematical definition of vorticity, $\underline{\xi}$, is

$$\underline{\xi} = \nabla \wedge \underline{v} = \text{curl } \underline{v} \qquad (2.35)$$

The so-called vorticity equation governs the manner in which vorticity is generated, convected and diffused through a moving fluid and this conservation-type equation is constructed mathematically by taking the curl of the momentum conservation equation (2.34).

For ease of illustration suppose we take the case where the motion is referred to an inertial frame of reference and there are no body forces. Taking the curl of equation (2.34) with this restriction leads, after some algebraic manipulation to

$$\frac{D\underline{\xi}}{Dt} = (\underline{\xi} \cdot \nabla) \underline{v} + \nu \nabla^2 \underline{\xi} \qquad (2.36)$$

The detailed derivation of equation (2.36) from equation (2.34) may be found in many standard texts on fluid mechanics, for example Lamb (1945). Physically equation (2.36) implies that vorticity is generated in the flow by the term $(\underline{\xi} \cdot \nabla)\underline{v}$ and subsequently convected and diffused through the flow by means of the terms $\dfrac{D\underline{\xi}}{Dt}$ and $\nu \nabla^2 \underline{\xi}$ respectively.

Let us now consider the vorticity equation when the motion of the fluid is referred to a translating and rotating reference frame. In this case the vorticity is measured relative to the frame as is the velocity field. For convenience suppose the following assumptions are made in order to highlight the salient features. There are no body forces acting, the fluid is unheated so that the density is independent of temperature, the density of the fluid is also independent of pressure so that the density is in effect constant. Suppose further that the reference frame is Cartesian and rotates with uniform angular velocity. Under these circumstances the momentum equation becomes, using equations (2.9) and (2.34),

$$\frac{D\underline{v}^*}{Dt} + 2(\underline{\omega} \wedge \underline{v}) + (\underline{\omega} \wedge (\underline{\omega} \wedge \underline{r})) + f_o = -\frac{1}{\rho} \nabla p + \nu \nabla^2 \underline{v} \qquad (2.37)$$

In many situations of practical importance (see examples cited in sections 2.3 and 2.4) the centripetal and the translational acceleration of the origin combine to form a conservative field which can be described via a scalar function, ϕ, such that

$$\underline{\omega} \wedge (\underline{\omega} \wedge \underline{r}) + f_o = \nabla \phi \qquad (2.38)$$

under which circumstances equation (2.37) becomes

$$\frac{D\underline{v}}{Dt} + 2(\underline{\omega} \wedge \underline{v}) = -\nabla(\frac{p}{\rho} + \phi) + \nu \nabla^2 \underline{v} \qquad (2.39)$$

The vorticity equation relative to the rotating reference frame now follows by taking the curl of equation (2.39) and noting that curl $(\nabla) = 0$ identically. Thus

$$\frac{D\underline{\xi}}{Dt} = (\underline{\xi} \cdot \nabla)\underline{v} + \nu \nabla^2 \underline{\xi} - 2\nabla \wedge (\underline{\omega} \wedge \underline{v}) \qquad (2.40)$$

Equation (2.40) demonstrates the following important results. The centripetal and translational acceleration terms vanish identically from the vorticity equation within the constraints of the assumptions made. In this respect these terms are purely hydrostatic with a constant density fluid very much analogous to the earth's gravitational field.

* Recall that the operator $\dfrac{\partial^2}{\partial t^2}$ in equation (2.9) has now been replaced by the total derivative operator $\dfrac{D}{Dt}$.

A term having its origin in the Coriolis acceleration does however now appear in the vorticity equation. This implies an additional generation term tending to create vorticity or relative rotation of the flow. Let us examine this term in more detail. Using standard vector identities we may write

$$\nabla \wedge (\underline{\omega} \wedge \underline{v}) = \underline{\omega}(\nabla \cdot \underline{v}) - (\nabla \cdot \underline{\omega})\underline{v} + (v \cdot \nabla)\underline{\omega} - (\underline{\omega} \cdot \nabla)\underline{v} \tag{2.41}$$

The continuity equation applied to an incompressible fluid implies that $\nabla \cdot \underline{v} = 0$. Also if $\underline{\omega}$ is a constant vector then $\nabla \cdot \underline{\omega} = 0$. Making use of these results permits the reduction of equation (2.41) to

$$\nabla \wedge (\omega \wedge \underline{v}) = - (\omega \cdot \nabla)\underline{v} \tag{2.42}$$

Combining equations (2.40) and (2.42) enables the vorticity equation to be simplified to

$$\frac{D\underline{\xi}}{Dt} = (\underline{\xi} \cdot \nabla)\underline{v} + \nu\nabla^2\underline{\xi} + 2(\underline{\omega} \cdot \nabla)\underline{v} \tag{2.43}$$

Equation (2.43) illustrates the important result that, provided $(\underline{\omega} \cdot \nabla)\underline{v}$ is non-zero, the Coriolis acceleration is capable of generating vorticity in a rotating frame of reference. This implies, for example, that Coriolis acceleration could under certain circumstances generate secondary flows in planes perpendicular to the main directions of flow for rotating ducts and that, in turn, this could significantly affect resistance to flow and convective heat or mass transfer inside the duct.

2.5.4 The Vorticity Equation for Heated Flow in a Rotating Duct.

The last sub-section demonstrated that, for constant property flow, the conservative combination of the centripetal acceleration and translational acceleration of the origin is hydrostatic in nature. In other words they do not contribute to the generation of vorticity. This is not the case when the flow is heated and a temperature-induced variation of fluid density is permitted. Under these circumstances vorticity may be generated due to a buoyancy-type effect as shown below.

If, as a result of heat transfer, the temperature field is non-uniform the local density, ρ, of the fluid may be related to the temperature, T, via an equation of state having the form

$$\rho = \rho_0\left[1 - \beta(T - T_0)\right] \tag{2.44}$$

where ρ_0 is the density of the fluid at a specified reference temperature T_0, and β is the coefficient of cubical expansion.

When the velocity field is zero and the fluid is at a uniform temperature, T_0, the hydrostatic pressure distribution, p_0, in the fluid is given by

$$\frac{P_o}{\rho_o} + \phi = 0 \qquad (2.45)$$

Suppose now that, with a particular heated flow, the pressure and fluid density are perturbed about the hydrostatic condition to give

$$P = P_o + P' \qquad (2.46)$$

and

$$\rho = \rho_o + \rho' \qquad (2.47)$$

where p' is the departure of the true pressure from the hydrostatic condition and ρ' is the density perturbation. Substitution of equations (2.46) and (2.47) into equation (2.39) yields

$$\frac{D\underline{v}}{Dt} + 2(\underline{\omega} \wedge \underline{v}) = - \frac{1}{(\rho_o + \rho')} \nabla (p_o + p') - \nabla\phi + \nu\nabla^2\underline{v} \qquad (2.48)$$

Expansion of the term $(\rho_o + \rho')$ using the Binomial theorem and neglecing second and higher order incremental terms enables equation (2.48) to be simplified to

$$\frac{D\underline{v}}{Dt} + 2(\underline{\omega} \wedge \underline{v}) = - \frac{1}{\rho_o} \nabla p' - \frac{\rho'}{\rho_o} \nabla\phi + \nu\nabla^2\underline{v} \qquad (2.49)$$

Implicit in equation (2.49) is the assumption that the kinematic viscosity, ν, remains constant even though changes in density are permitted. The density perturbation can be written in terms of the temperature of the fluid using equation (2.44) so that equation (2.49) becomes

$$\frac{D\underline{v}}{Dt} + 2(\underline{\omega} \wedge \underline{v}) = - \frac{1}{\rho_o} \nabla p' + \beta(T - T_o)\nabla\phi + \nu\nabla^2\underline{v} \qquad (2.50)$$

If the scalar function, ϕ, is thought of as an effective body force then implicit in equation (2.50) is the usual Boussinesq (1930) approximation that the density variation need be included only in the effective body force term. This is important since to a first approximation the temperature-density variation need not be included in the Coriolis term.

Equation (2.50) can also be transformed into an equivalent vorticity equation which becomes

$$\frac{D\underline{\xi}}{Dt} = (\underline{\xi}\cdot\nabla)\underline{v} + \nu\nabla^2\underline{\xi} + 2(\underline{\omega}\cdot\nabla)\underline{v} + \nabla\wedge(\beta(T - T_o)\nabla\phi) \qquad (2.51)$$

Equation (2.51) permits the following observations for a heated rotating duct. Coriolis acceleration generates vorticity as was shown earlier provided $(\underline{\omega}\cdot\nabla)\underline{v}$ is non-zero and temperature-induced density variations do not directly affect this term to a first order approximation. The conservative effective body force yield described by the scalar function ϕ can create vorticity via a temperature density interaction. Thus in a heated rotating tube the flow field will be simultaneously influenced by Coriolis acceleration and a centripetal-type buoyancy. Clearly the buoyancy effect will depend markedly on the relative location of the tube to the axis of rotation and also on the direction of the main through flow. These overall effects of rotation will now be discussed in detail for a number of specific flow geometries having practical relevance.

CHAPTER 3

LAMINAR FLOW AND HEAT TRANSFER

IN CIRCULAR-SECTIONED TUBES WHICH

ROTATE ABOUT A PARALLEL AXIS

3.1 Introduction

In this chapter an appraisal of the current state of knowledge for the determination of flow and heat transfer characteristics of fluids moving in straight circular-sectioned ducts will be presented for the case where the duct is constrained to rotate about a parallel axis and the flow is laminar. This class of rotating flow geometry often features in the design of cooling systems for electrical machine rotors as was illustrated in Chapter 1 and is accordingly of significant practical importance. The chapter commences with a study of isothermal flow and subsequently leads on to the treatment of heated flows. The corresponding problem for turbulent flow will be studied in Chapter 4.

3.2 Isothermal Flow in Circular-Sectioned Ducts.

Within the limitations of the assumptions made, it was demonstrated in the previous chapter how Coriolis forces could induce vorticity or secondary flow in a rotating duct whereas the conservative centripetal type terms were mainly hydrostatic when the flow was isothermal implying in this context, that the fluid has constant density. Let us now develop these ideas and their consequences for the flow systems considered in this chapter.

Consider the case of laminar flow with the motion referred to a Cartesian frame as studied in section 2.3 and illustrated schematically in figure 2.2. Taking the acceleration vector for a typical fluid particle from equation (2.15) and inserting into equation (2.39) permits the momentum equations for the x, y and z directions respectively to be determined as

$$\frac{Du}{Dt} - 2\Omega v = - \frac{1}{\rho} \frac{\partial p}{\partial x} + \Omega^2 (H + x) + \nu \nabla^2 u$$

$$\frac{Dv}{Dt} + 2\Omega u = - \frac{1}{\rho} \frac{\partial p}{\partial y} + \Omega^2 y + \nu \nabla^2 v$$

$$\frac{Dw}{Dt} = -\frac{1}{\rho}\frac{\partial p}{\partial z} + \nu\nabla^2 w \qquad \qquad \qquad (3.1)$$

The assumptions made in sub-section 2.5.3 are applicable in this instance so that

$$\phi = -\Omega^2(Hx + \frac{1}{2}(x^2 + y^2)) \qquad (3.2)$$

Also, since $\underline{\omega} = \Omega\underline{k}$ and $\underline{v} = u\underline{i} + v\underline{j} + w\underline{k}$, the Coriolis-induced vorticity term $2(\underline{\omega}\cdot\nabla)\underline{v}$ of equation (2.43) becomes

$$2(\underline{\omega}\cdot\nabla)\underline{v} = 2\Omega\left[\frac{\partial u}{\partial z}\underline{i} + \frac{\partial v}{\partial z}\underline{j} + \frac{\partial w}{\partial z}\underline{k}\right] \qquad (3.3)$$

This is an interesting result since it implies that the vorticity generation due to Coriolis forces will vanish identically in this case provided fully developed flow has been established. Thus when there are no axial velocity gradients the flow field will be unchanged in relation to the stationary case and the influence of rotation will be made manifest as a cross stream radial equilibrium pressure field in the XY-plane. The pressure field will consequently have the form

$$p = \gamma z + \rho\Omega^2(Hx + \frac{1}{2}(x^2 + y^2)) \qquad (3.4)$$

where γ is a constant axial pressure gradient which results from an appropriate stationary duct solution with established flow.

Morris (1965a) developed a similar result for developed laminar flow in a tube having a circular cross section. Although, as pointed out in sub-section 2.3, the use of a cylindrical polar reference frame implies an angular acceleration of the reference frame, the same conclusions concerning the general effect of Coriolis and centripetal-type accelerations emerge. Hence for developed flow in a circular sectioned tube with the flow system shown in figure 2.3 the pressure distribution has the form

$$p = \gamma z + \rho\Omega^2(H\cos\theta + \frac{1}{2}r)r \qquad (3.5)$$

In effect, therefore, no secondary flow occurs in the established regime and a cross stream forced vortex-type radial equilibrium pressure field is the manifestation of rotation. Under these circumstances the axial pressure gradient would be calculated from the usual Blasius friction factor, C_f, where

$$C_f = \frac{64}{Re} \qquad (3.6)$$

and Re is the usual through flow Reynolds number.

The above results are only applicable for developed laminar flow. Equation (3.3) clearly shows that in the entrance region of tubes

rotating in the parallel mode there will be a strong tendency to generate relative vorticity and this must have a consequential effect on
the resistance offered to flow. Intuitively one expects the additional mixing implied by axial gradients of velocity to modify the axial
velocity field and probably increase the resistance offered to flow.
The only experimental data presently available in this respect is a
recent study by Morris (1981). Although the experimental data treated a notionally turbulent flow regime some of the trends reported
highlight some important new features and these will now be discussed
in some detail.

Figure 3.1 shows the scantlings of the flow circuit used for this
experimental study of flow resistance. In essence the apparatus consisted of a built up rotor which supported a test section comprising
a circular tube 9.5 mm bore diameter by 610.0 mm in length rotating
with an eccentricity of 457.2 mm. Internal passages permitted an air
flow to and from the test section via rotating seal assemblies. The
test section itself was fitted with pressure tappings at quarter span
intervals along its length and the pressure signals were transmitted
hydrostatically from the rotor to stationary pressure measuring inst-
ruments using a specially designed pressure transmission system in-
corporating magnetically sealed chambers. An initial programme of
pressure drop measurements demonstrated that the equipment functioned
satisfactorily at zero rotational speed. Tests under conditions of
rotation were performed at speeds upto 600 rev/min with nominal
through flow Reynolds numbers in the range 2500 - 10,000.

Figure 3.2 typifies the manner in which rotation was found to influence resistance to flow. Here, for a selection of Reynolds number
values, are shown the Blasius friction factors measured at zero rota-
ticnal speed and also at 600 rev/min evaluated at quarter span intervals along the test section (i.e. length/diameter ratios of 16, 32,
48 and 64 respectively). An interesting behavioural pattern may be
seen. At the lowest Reynolds number (curve A of figure 3.2) where
the flow was nearest to the conventional laminar regime, rotation
produced significant increases in friction factor. This, of course,
could be expected in view of the earlier comments concerning Coriolis-
induced secondary flows in the entrance region. However as the flow
rate was increased there was a tendency for the relative increase in
friction factor to decrease until, at a Reynolds number value of 7396
(curve C in figure 3.2), the results at both speeds were virtually
indistinguishable. An increase in flow rate beyond this value reversed the trend in which case rotation was found to bring about a significant reduction in friction factor (see curve D of figure 3.2).

An explanation for these trends was postulated in terms of a possible delay in the transition from laminar to turbulent flow due to some
stabilizing influence of rotation. Justification for this hypothesis
is evident on reference to figure 3.3. Here the friction factor results obtained at 300 and 600 rev/min are plotted and compared with
the usual stationary tube correlations. The data for $\frac{L}{d}$ = 64 is actually shown. With laminar flow in stationary tubes, the friction
factor is inversely proportional to the Reynolds number. The data
points presented by Morris (1981) at the two speeds shown in figure

36

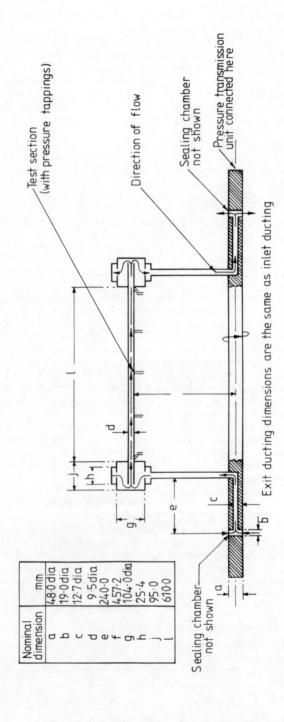

Nominal dimension	mm
a	48·0 dia
b	19·0 dia
c	12·7 dia
d	9·5 dia
e	240·0
f	457·2
g	104·0 dia
h	25·4
j	95·0
l	610·0

Test section (with pressure tappings)

Direction of flow

Sealing chamber not shown

Pressure transmission unit connected here

Sealing chamber not shown

Exit ducting dimensions are the same as inlet ducting

FIG. 3.1   LEADING DIMENSIONS OF FLOW CIRCUIT STUDIED BY MORRIS (1981) FOR FLOW RESISTANCE MEASUREMENTS

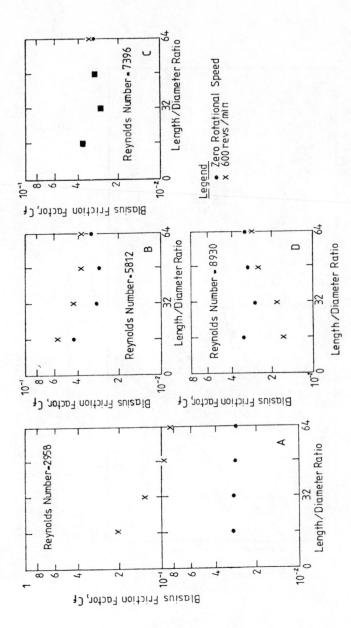

FIG. 3.2 TYPICAL EFFECT OF ROTATION ON BLASIUS FRICTION FACTOR FOR A RANGE OF REYNOLDS NUMBERS AND LENGTH/DIAMETER RATIOS. MORRIS (1981).

38

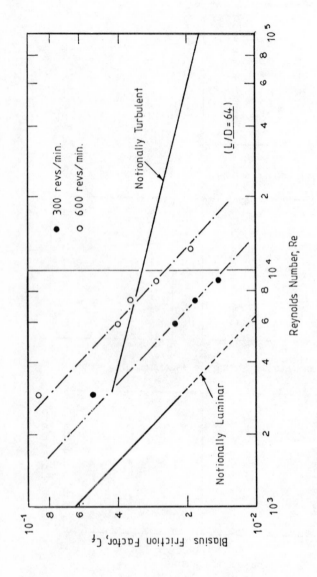

FIG. 3.3   TYPICAL INFLUENCE OF ROTATION ON THE BLASIUS FRICTION FACTOR. MORRIS (1981)

3.3 had a strong tendency to follow such a relationship even though the Reynolds number values were well into the customary turbulent range. Indeed all tests undertaken in the overall speed range 0 - 600 rev/min were found to follow this trend.

The pattern of results depicted in figure 3.3 suggested the possibility of correlating the rotating friction factor data by an equation having the mathematical structure

$$C_{fR} = C_{fo} + C_f^*$$ (3.7)

where C_{fR} is the friction factor obtained with rotation and a specified Reynolds number and flow geometry. C_{fo} is the friction factor determined by assuming the flow to be laminar at the same specified Reynolds number and calculated using equation (3.6). C_f^* was postulated as a so-called "excess friction factor" which relates C_{fR} and C_{fo}. Morris (1981) used the following dimensional arguments to seek a provisional empirical correlation for the excess friction factor.

Using the acceleration components determined for the parallel mode of rotation and a polar reference frame from equation (2.21), the laminar momentum conservation equations become, for the r, θ, and z directions respectively

$$\frac{Du}{Dt} - \frac{v^2}{r} - 2\Omega v = -\frac{1}{\rho}\frac{\partial p}{\partial r} + \Omega^2(H\cos\theta + r) + \nu\left[\nabla^2 u - \frac{u}{r^2} - \frac{2}{r^2}\frac{\partial v}{\partial\theta}\right]$$

$$\frac{Dv}{Dt} + \frac{uv}{r} + 2\Omega u = -\frac{1}{\rho r}\frac{\partial p}{\partial\theta} - \Omega^2 H\sin\theta + \nu\left[\nabla^2 v - \frac{v}{r^2} + \frac{2}{r^2}\frac{\partial u}{\partial\theta}\right] \quad (3.8)$$

$$\frac{Dw}{Dt} = -\frac{1}{\rho}\frac{\partial p}{\partial z} + \nu\nabla^2 w$$

The non-dimensional groups likely to control the flow field maybe determined by making the transformations

$$U = \frac{u}{w_m}, \quad V = \frac{v}{w_m}, \quad W = \frac{w}{w_m}$$ (3.9)

$$R = \frac{r}{d}, \quad Z = \frac{z}{d}$$ (3.10)

$$P = \frac{p}{\rho w_m^2}$$ (3.11)

where w_m is the mean axial velocity of the fluid in the tube and d is the diameter of the tube. Substituting equations (3.9) through (3.11) into the radial momentum equation by way of example gives

$$\frac{DU}{Dt} - 2Ro\,V = -\frac{\partial P}{\partial R} + Ro^2(\epsilon\cos\theta + R) + \frac{1}{Re}\left[\nabla^2 U - \frac{U}{R^2} - \frac{2}{R^2}\frac{\partial V}{\partial\theta}\right] \qquad (3.12)$$

where

$$Ro = \frac{\Omega d}{w_m} \qquad \text{(Rossby number)}$$

$$Re = \frac{w_m d}{\nu} \qquad \text{(Reynolds number)} \qquad (3.13)$$

$$\epsilon = \frac{H}{d} \qquad \text{(Eccentricity parameter)}$$

Equation (3.12) demonstrates that the flow will be parametrically governed by three non-dimensional groups. As well as the usual Reynolds number it is necessary to include the eccentricity parameter and the so-called Rossby number. The Rossby number has its origin in the study of Coriolis-induced secondary flows in the earth's atmosphere and maybe thought of as a ratio of Coriolis to inertial forces acting. Morris (1981) examined his pressure drop data, which were taken at a fixed value of the eccentricity parameter, to see if the excess friction factor could be correlated in terms of the Rossby number. To this end C_f^* was evaluated at all speeds in the range 0-600 rev/min and figure 3.4 shows the trends noted. All data points showed a strong tendency to collapse onto a unique line which, Morris (1981) proposed, could be empirically represented by

$$C_f^* = 0.503\,Ro^{1.06} \qquad (3.14)$$

Although these results are limited in the range of variables studied they represent virtually the only direct measurements currently available.

To summarise, it is apparent with isothermal incompressible flow that rotation will have no effect on flow resistance provided the flow is laminar and developed. For developing laminar flow Coriolis effects will produce cross stream secondary flow and an attendent increase in friction factor. Under these circumstances equation (3.14) may be used tentatively to give some indication of the magnitude of the expected effect. Since the eccentricity parameter emerges from the centripetal-type terms it is likely to have a small order influence as a provisional estimate.

For turbulent flow the stabilising effect of rotation on the transition from laminar to turbulent flow suggests that, in overall terms, a reduction in friction factor may be experienced when rotation is considered. Again within the limitation of the range of variables experimentally tested by Morris (1981), equation (3.14) may be used for a tentative appraisal of pressure loss for circular pipes. Further design implications will be given at the end of this chapter.

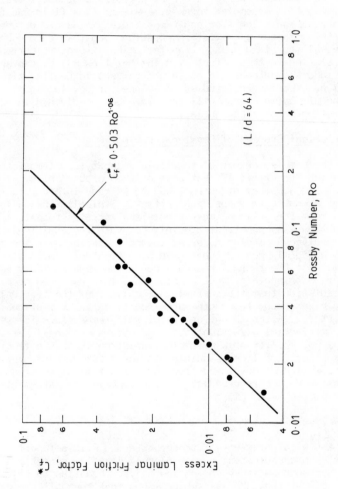

FIG. 3.4 COMPARISON OF EXCESS LAMINAR FRICTION FACTOR WITH ROSSBY NUMBER FOR ALL SPEEDS TESTED. MORRIS (1981).

3.3   Heated Flow in Circular-Sectioned Ducts

3.3.1   Opening Remarks

We now turn our attention to a discussion of flow and heat transfer
with parallel-mode rotation.   With heated flow it was shown in Chapter 2 that the simultaneous influence of Coriolis and centripetal
forces can result in a complex three dimensional flow field and temperature distribution.   The flow and temperature fields are directly
coupled in the momentum conservation equations and this consequently
affects the wall to fluid energy transfer.   The literature currently
available relating to this effect with laminar flow may be conveniently sub-divided and categorised as to whether one is dealing with
theoretical or experimental studies, developed or developing flow, etc.
This section of the monograph will therefore be sub-divided in a manner which reflects these categories.

3.3.2   Theoretical Studies of Developed Laminar Flow

Owing to the complex nature of the fluid velocity and temperature
fields existing, theoretical treatments of this problem have, of necessity, required a number of assumptions to be made in order to render
the problem amenable to theoretical attack.   Typically the following
assumptions form the basis of most theoretical studies to date.
Developed flow has been treated mainly to avoid the additional dif-
ficulties involved by including axial velocity gradients in the analysis.   Also uniformly heated tubes have been treated.   In this respect
the uniformity of the thermal boundary condition means uniform heat
transfer per unit length of duct considered.   This has been coupled
with the assumption that the thermal conductivity of the tube wall
material is high enough to smooth out circumferential variations in
wall temperature.   Although uniformity of wall temperature is assumed
in the circumferential direction this does not, as we shall see later,
preclude the possibility of a heat flux variation in this direction
even with constant total heat addition in the axial direction.   The
combined assumption of developed flow and uniform axial heating con-
strains the fluid temperature distribution to have the mathematical
structure

$$T = \tau z + F(r, \theta) \qquad\qquad (3.15)$$

where τ is the axial gradient of temperature and F is a function of
the cross stream coordinates r and θ alone.   The flow geometry and
coordinate system used is shown in figure 3.5.   Equation (3.15) is
also applicable at the tube wall which means that the wall temperature
will increase uniformly in the direction of flow.   Further, at any
axial location the difference in the wall temperature, T_w, and any
local value of temperature in the flow will also be functionally related to cross stream coordinates alone.   This is the so-called similarity solution for the temperature field.
The first attempt to theoretically study this problem was made by
Morris (1965b).   In his investigation the axis of rotation was taken

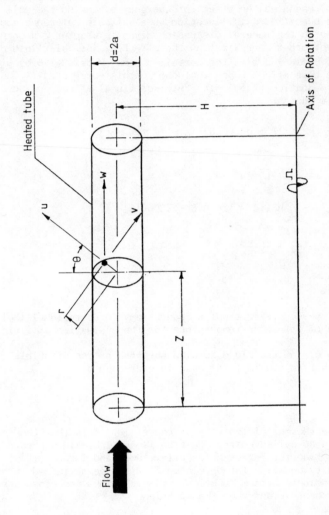

FIG. 3.5   ROTATING GEOMETRY COORDINATE SYSTEM FOR CIRCULAR-SECTIONED TUBE.

to be vertical and the additional influence of the earth's gravitational buoyancy included in the axial direction. This effect will be omitted in the review presented here in the interest of clarity.

Because of the simplification implied by the temperature similarity Morris (1965b) measured the effect of buoyancy by taking the wall temperature at a specified axial location as a suitable reference condition. Although the general discussion given in Chapter 2 suggests that to a first order approximation the effect of variable fluid density need not be included in the Coriolis terms, Morris (1965b) did indeed permit this effect to be included in this early study. Thus evaluation of equation (2.50) with this additional effect included gives for this geometry

$$\frac{Du}{Dt} = -\frac{1}{\rho}\frac{\partial p'}{\partial r} + \Omega^2\beta(r + H\cos\theta)(T_w - T) + 2\Omega\beta v(T_w - T) +$$

$$+ \nu(\nabla^2 u - \frac{u}{r^2} - \frac{2}{r^2}\frac{\partial v}{\partial\theta}) \tag{3.16}$$

$$\frac{Dv}{Dt} = -\frac{1}{\rho r}\frac{\partial p'}{\partial\theta} - \Omega^2 H\beta(T_w - T)\sin\theta - 2\Omega\beta u(T_w - T) +$$

$$+ \nu(\nabla^2 v - \frac{v}{r^2} + \frac{2}{r^2}\frac{\partial u}{\partial\theta}) \tag{3.17}$$

$$\frac{Dw}{Dt} = -\frac{1}{\rho}\frac{\partial p'}{\partial z} + \nu\nabla^2 w \tag{3.18}$$

Here p' is a pseudo pressure which absorbs any force residual implied by the choice of reference temperature for the evaluation of fluid density.

The temperature of the fluid is also governed by the principle of energy conservation and in this case is expressed by

$$\frac{DT}{Dt} = \alpha\nabla^2 T \tag{3.19}$$

where α is the thermal diffusivity of the fluid. Note that the derivation of this so-called energy equation is available in standard texts on heat transfer for example Bayley, Owen and Turner (1972).

The continuity equation for developed flow may be satisfied by means of a two-dimensional stream function, Ψ, which links the cross stream velocity components u and v by the equation

$$\frac{\partial\Psi}{\partial r} = -\frac{v}{\nu} \quad \text{and} \quad \frac{\partial\Psi}{\partial\theta} = \frac{ru}{\nu} \tag{3.20}$$

The stream function, by definition, is linked to the vorticity function, ξ, by

$$\nabla^2\Psi = -\xi \tag{3.21}$$

Because the flow is developed the pseudo pressure field is constrained to have the form

$$p' = \gamma z + G(r,\theta) \tag{3.22}$$

where γ is a constant axial gradient of pseudo pressure and G is a function of cross stream coordinates alone.

Morris (1965b) recast equations (3.16) through (3.19) in non-dimensional form and introduced the stream function to replace components of velocity in the cross stream directions. The non-dimensionalising scheme adopted for the dependent and independent variables was

$$R = \frac{r}{a} \quad , \quad Z = \frac{z}{a} \tag{3.23}$$

$$W = \frac{wa}{\nu} \quad , \quad \eta = \frac{(T_w - T)}{\tau a Pr} \tag{3.24}$$

where Pr is the Prandtl number of the fluid and a is the radius of the tube. Using these transformations and eliminating the pressure terms from the radial and tangential momentum equations by cross differentiation and subtraction it is possible to generate the following set of equations.

$$\nabla^4 \Psi + \frac{1}{R} \frac{\partial(\Psi, \nabla^2 \Psi)}{\partial(R, \theta)} + Ra_\tau \left[\frac{1}{R} \frac{\partial \eta}{\partial \theta} \cos\theta + \frac{\partial \eta}{\partial R} \sin\theta \right]$$

$$+ \frac{Ra_\tau}{\varepsilon_a} \frac{\partial \eta}{\partial \theta} + \frac{Ra_\tau \cdot Ro'}{Re_p} \frac{\partial(\eta, \Psi)}{\partial(R, \theta)} = 0 \tag{3.25}$$

$$\nabla^2 W + \frac{1}{R} \frac{\partial(\Psi, W)}{\partial(R, \theta)} + 4 Re_p = 0 \tag{3.26}$$

$$\nabla^2 \eta + \frac{Pr}{R} \frac{\partial(\Psi, \eta)}{\partial(R, \theta)} + W = 0 \tag{3.27}$$

where the Jacobian operator is defined as

$$\frac{\partial(A, B)}{\partial(R, \theta)} = \frac{\partial A}{\partial R} \frac{\partial B}{\partial \theta} - \frac{\partial B}{\partial R} \frac{\partial A}{\partial \theta} \tag{3.28}$$

The non-dimensionalisation procedure used highlights the following non-dimensional groups which parametrically govern this problem

$$Ra_\tau = \frac{\Omega^2 H \beta \tau a^4}{\alpha \nu^2} \qquad \text{(Rotational Rayleigh Number)}$$

$$Ro' = \frac{-a^2}{2H\Omega\rho\nu} \frac{\partial p'}{\partial z} \qquad \text{(A form of the Rossby Number)}$$

$$Pr = \frac{\nu}{\alpha} \qquad \text{(Prandtl Number)}$$

$$Re_p = \frac{-a^3}{\rho \nu^2} \frac{\partial p'}{\partial z} \qquad \text{(Pseudo Reynolds Number)} \qquad (3.29)$$

$$\varepsilon_a = \frac{H}{a} \qquad \text{(Eccentricity Parameter)}$$

The rotational Rayleigh number emerges from the centripetal buoyancy terms of the momentum equations. This is similar to the Rayleigh number encountered in the study of gravitational buoyancy due to the earth's field but with the gravitational acceleration replaced by the centripetal acceleration measured at the centre line of the tube considered. The Rossby number, Ro', has its origin in the Coriolis terms and may be thought of as an alternative to the Rossby number, Ro, used in Chapter 2 for the description of the Coriolis importance. The pseudo Reynolds number, Re_p, is defined in an identical mathematical form to the usual through flow Reynolds number, Re, and to which it reduces when buoyancy effects are not included in the analysis.

Equations (3.25) through (3.27) must be solved subject to the usual zero slip conditions at the wall where by definition $(T_W - T)$ is also zero. The velocity and temperature must also be finite at the centre line of the tube. Thus the solution must satisfy the boundary conditions

$$W = \frac{\partial \Psi}{\partial R} = \frac{\partial \Psi}{\partial \theta} = \eta = 0 \text{ at } R = 1 \qquad (3.30)$$

In order to solve equations (3.25) through (3.27) Morris (1965b) used a series expansion technique proposed by Lighthill (1949). Applied to this problem the technique involved expansion of the non-dimensional velocity and temperature field in ascending powers of a suitably small parameter. The method has previously been used by Morton (1959) for a study of gravitational buoyancy in heated horizontal tubes with forced laminar flow. Morris (1965b) selected the Rayleigh number as the parameter with which to form the series expansion solution. Thus it was assumed that the stream function, axial velocity and temperature distribution could be expressed as

$$\Psi = \Psi_0 + Ra_\tau \Psi_1 + Ra_\tau^2 \Psi_2 \qquad (3.31)$$

$$W = W_0 + Ra_\tau W_1 + Ra_\tau^2 W_2 \qquad (3.31)$$

$$\eta = \eta_0 + Ra_\tau \eta_1 + Ra_\tau^2 \eta_2 \qquad (3.33)$$

Unfortunately the requirement that Ra_τ is small in order that this method of solution is applicable restricts the solution to low rotational speeds and heating rates. In this respect the results serve only to give an assessment of the likely physical manifestations of rotation.

On substitution of equations (3.31) through (3.33) into the conservation equations, sets of zero, first and second order equations are obtained by equating powers of the rotational Rayleigh number. Each set of equations may be solved sequentially commencing at that of zero order. Morris (1965b) solved upto and including the second order terms. Because the resulting algebra is extremely tedious only details for the zero and first order solutions are presented here.

Axial Velocity

$$
\left.
\begin{aligned}
&W_0 = Re_p(1 - R) \\[2mm]
&W_1 = Re_p^2\, R(1 - R^2)(49 - 51R^2 + 19R^4 - R^6)\cos\theta\, / \\[2mm]
&\qquad /1.843 \times 10^5
\end{aligned}
\right\} \quad (3.34)
$$

Stream Function

$$
\left.
\begin{aligned}
&\Psi_0 = 0 \\[2mm]
&\Psi_1 = Re_p\, R(1 - R^2)(10 - R^2)\sin\theta\,/4608
\end{aligned}
\right\} \quad (3.35)
$$

Temperature

$$
\left.
\begin{aligned}
&\eta_0 = Re_p\,(1 - R^2)(3 - R^2)\,/16 \\[3mm]
&\eta_1 = Re_p^2[(381 + 1325Pr)R - (175 + 3000Pr)\,R^3 + \\[2mm]
&\qquad + (500 + 2600Pr)R^5 - (175 + 1125Pr)R^7 + \\[2mm]
&\qquad + (30 + 210Pr)R^9 - (1 + 10Pr)R^{11}]\cos\theta\,/ \\[2mm]
&\qquad /2.217 \times 10^7
\end{aligned}
\right\} \quad (3.36)
$$

Morris (1965b) examined the influence of rotation on the established resistance to flow by evaluating a Blasius friction factor, C_{fR} defined as

$$
C_{fR} = \frac{-a}{\rho w_m^2}\,\frac{\partial p'}{\partial z} \quad (3.37)
$$

where the mean axial velocity, w_m, was determined from integration of the axial velocity profile to give

$$
w_m = \frac{2\,Re_p\upsilon}{a}\left\{\frac{1}{4} - 0.0525\left[\frac{Re_p\,Ra_T}{4608}\right]^2\right\} \quad (3.38)
$$

Note also that equation (3.38) gives a direct link between the pseudo and through flow Reynolds numbers as

$$Re = \frac{w_m d}{\upsilon} = Re_p\left(1 - 0.21\left[\frac{Re_p\ Ra_\tau}{4608}\right]^2\right) \qquad (3.39)$$

Making use of equations (3.37) through (3.39) permits the following expression for the Blasius friction factor to be determined

$$C_{fR} = \frac{64}{Re_p\left(1 - 0.21\left[\frac{Re_p\ Ra_\tau}{4608}\right]^2\right)^2} = \frac{64\ Re_p}{Re^2} \qquad (3.40)$$

Alternatively, using equation (3.6), the ratio of the rotating to non-rotating friction factor becomes

$$\frac{C_{fR}}{C_{fo}} = \frac{1}{\left(1 - 0.21\left[\frac{Re_p\ Ra_\tau}{4608}\right]^2\right)} \qquad (3.41)$$

Figure 3.6 shows the way in which rotation affects the axial velocity distribution for solutions upto first order. Here for $Re_p = 100$, $Pr = 1$ and a range of Rayleigh number values, we see how the axial velocity profile distorts so that the location of the point of maximus velocity moves away from the axis of rotation. Note that, upto second order solutions, the Rossby number has no effect on this profile. The effect of rotation on the friction factor is shown in figure 3.7. Here the ratio of the rotating to non-rotating friction factor is plotted against the product of the pseudo Reynolds and Rayleigh numbers. This product is a naturally emerging group of nondimensional variables from the analysis. The increased resistance to flow resulting from the cross stream secondary flow is clearly demonstrated.

An alternative illustration of the way in which the combined influence of rotation and heating affects flow resistance is given in figure 3.8. For a range of rotational Rayleigh numbers the implied links between the through flow Reynolds number and the pseudo Reynolds number is shown. If we think of the pseudo Reynolds number as a measure of prescribed pressure gradient and the through flow Reynolds number as a measure of flow then figure 3.8 clearly demonstrates the manner in which flow is reduced for a specified pressure gradient as the rotational Rayleigh number increases. The rapid fall off in flow rate for $Ra_\tau = 100$ and $Re_p > 80$ is probably indicative that the validity of the series expansion technique is being violated. Nevertheless the reduction in flow rate which physically occurs as the rotational buoyancy increases is certainly demonstrated.

For the same operating conditions as those specified for the axial velocity profile shown in figure 3.6, the typical effect of rotation

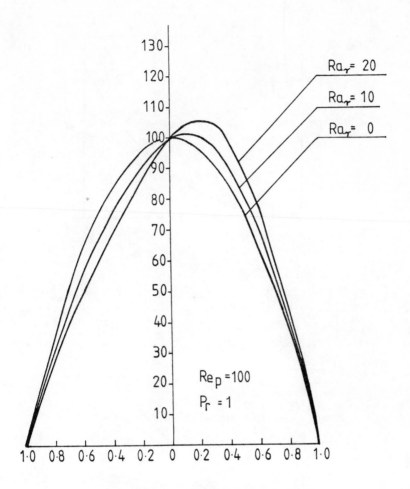

FIG. 3.6   TYPICAL EFFECT OF ROTATION ON AXIAL
            VELOCITY PROFILE UP TO FIRST ORDER
            SOLUTION. MORRIS (1965b).

on the temperature distribution is shown in figure 3.9.   The tendency
for the cooler and less dense fluid to move towards the outer region
of the tube cross section is clearly shown.   The consequential effect
of the temperature field distortion on heat transfer from the wall of
the tube to the fluid may be determined as follows.

At the wall of the tube heat transfer to the fluid is solely due to
conduction in the radial direction.   Thus for a small angular element
of tube periphery, the locally prevailing heat transfer rate, $\dot{Q}(\theta)$
may be expressed as

50

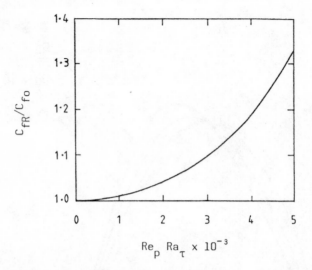

FIG. 3.7   INFLUENCE OF ROTATION ON FLOW RESISTANCE
            MORRIS (1965b).

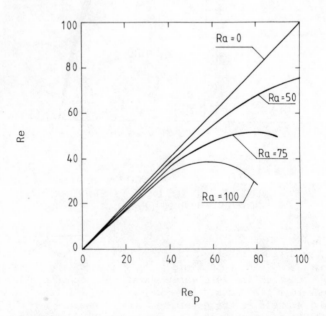

FIG. 3.8   EFFECT OF ROTATION ON MEAN FLOW RATE
            MORRIS (1965b).

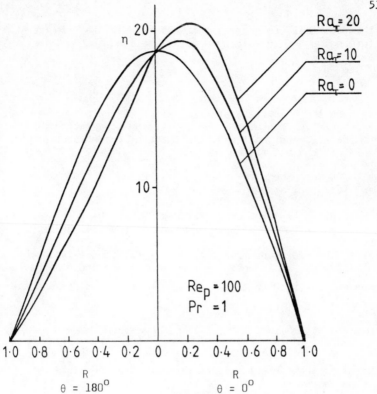

FIG. 3.9   TYPICAL EFFECT OF ROTATION ON TEMPERATURE
PROFILE UPTO FIRST ORDER SOLUTION.
MORRIS (1965b).

$$\dot{Q}(\theta) = - ka \left. \frac{\partial T}{\partial r} \right]_{r=a} \delta\theta \qquad\qquad (3.42)$$

where k is the thermal conductivity of the fluid.

Expressing equation (3.42) in terms of the non-dimensional dependent and independent variables, given by equations (3.23) and (3.24)
respectively, and integrating over the circumference of the tube enables the average heat transfer rate, $\dot{Q}$, to be expressed as

$$\dot{Q} = ka_T Pr \int_0^{2\pi} \left. \frac{\partial \eta}{\partial R} \right]_{R=1} d\theta \qquad\qquad (3.43)$$

Note that our original thermal boundary condition postulated that $\dot{Q}$
should be uniform in the direction of flow. Since the temperature
profile is asymmetric, this does not preclude a girthwise variation
of heat transfer rate at the tube-fluid interface however.

It is customary to express the heat transfer rate in non-dimensional form using the Nusselt number, Nu, defined as

$$Nu = \frac{\dot{q}\, d}{\Delta T_w k} \qquad (3.44)$$

where $\dot{q}$ is the heat transfer rate per unit area of surface and ΔT_w is a representative measure of the motivating temperature difference between the wall and the fluid. In terms of the present problem equation (3.44) can be written as

$$Nu = \frac{\dot{Q}}{\pi k \Delta T_w} \qquad (3.45)$$

The motivating temperature difference ΔT_w for heat transfer may be expressed in a number of ways and two methods will be briefly discussed here.

First, the representative temperature difference is taken to be $(T_w - T_m)$ where T_m is an integrated mean temperature across the duct (sometimes referred to as an unweighted mean) and defined as

$$T_m = \frac{4}{\pi d^2} \int_0^{2\pi} \int_0^a T\, r\, dr\, d\theta \qquad (3.46)$$

In terms of the present problem studied it is possible to express equation (3.46) as

$$T_m = \frac{1}{\pi} \int_0^{2\pi} \int_0^1 (T_w - \tau a\, Pr\, \eta) R\, dR\, d\theta \qquad (3.47)$$

To experimentally determine the unweighted mean temperature it is necessary to know the variation of fluid temperature across the duct. This is often not possible to obtain without refined instrumentation and it is consequently more convenient to use a weighted mean temperature, T_b, (sometimes referred to as the bulk temperature) defined as

$$T_b = \frac{4}{\pi d^2 w_m} \int_0^{2\pi} \int_0^a w\, T\, r\, dr\, d\theta \qquad (3.48)$$

or, in terms of the present problem,

$$T_b = \frac{2\upsilon}{\pi d} \int_0^{2\pi} \int_0^1 W\, (T_w - \tau a\, Pr\, \eta) R\, dR\, d\theta \qquad (3.49)$$

Thus an alternative representation of the motivating temperature difference may be taken as $(T_w - T_b)$.

Morris (1965b) elected to use the unweighted mean temperature of the fluid to define the motivating temperature difference for heat transfer due to the extremely tedious algebra needed to integrate the product of the axial velocity field and the temperature field. The corresponding Nusselt number $Nu_{\infty,u}$ was subsequently evaluated as

$$
Nu_{\infty,u} = \frac{0.2500 - \left[\dfrac{Re_p Ra_\tau}{4608}\right]^2 \left[0.0328 + 0.0000Pr + 0.0018Pr^2\right]}{0.0417 - \left[\dfrac{Re_p Ra_\tau}{4608}\right]^2 \left[0.0133 + 0.0035Pr + 0.0009Pr^2\right]}
\tag{3.50}
$$

where numerical constants are given to four significant figures.

The following implications of equation (3.50) are worthy of note. In the limiting case of zero rotation the present problem becomes identical to the classic Nusselt (1910) problem for constant property forced convection and reported by Goldstein (1957). For this limited case $Nu_o = 6$ and clearly equation (3.50) asymptotes to this value as $Ra_\tau \to 0$.

The series expansion solution method was evaluated by Morris (1965b) upto and including second order solutions. Although the Rossby number, Ro' was present in the second order solutions and also the eccentricity parameter, ε_a , they do not appear explicitly in the Nusselt Number expression. Within the limitations of the assumptions made the Nusselt number depends only on the Prandtl number and the product of the pseudo Reynolds number and rotational Rayleigh number.

Figure 3.10 illustrates typical trends for the Nusselt number variation with rotation and Prandtl number. Although it is likely that the solution is being extended beyond its range of validity at the higher values of the $Re_p Ra_\tau$ product shown the physical trends are clearly evident. At a specified Prandtl number value the heat transfer progressively increases as the product of $Re_p Ra$ increases. At a specified value of the $Re_p Ra$ product we note the important fact from figure 3.10 that the heat transfer enhancement is relatively greater at the higher Prandtl number values. Thus, for example, the relative increase in Nusselt number will be higher with liquid-like flows in comparison to gas-like flows.

The series expansion technique described above restricts the validity of the solution to low values of the rotational Rayleigh number. This implies low rates of heating and rotation in real terms. It is not possible to state with certainty the quantitative limits beyond which the solution is invalid. Even so we can observe, from figure 3.10 for example, the tendency for the solution to "run away" as the rotational Rayleigh number is progressively increased. As a consequence it may be argued that the original analysis of Morris (1965b) was only useful for demonstrating the likely physical manifestations of rotation on the flow and heat transfer but was severely restricted for quantitative assessment.

To overcome this quantitative deficiency Mori and Nakayama (1967)

54

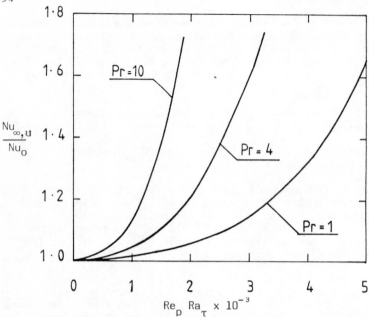

$$\frac{Nu_{\infty,u}}{Nu_0}$$

$Re_p \, Ra_\tau \times 10^{-3}$

FIG. 3.10   INFLUENCE OF ROTATION ON HEAT TRANSFER
MORRIS (1965b)

attempted to solve the same problem using, in effect, an integral-
type method which they argued was valid at high rotation speeds.   The
salient features of this method and the resulting solutions will now
be discussed.   Note that, in the interest of consistency, the original
nomenclature used by Mori and Nakayama (1967) has been changed to that
used in the earlier sections of this book.

At high rotational speeds it is likely that the strong cross stream
secondary flow generated will dominate shear stress and heat transfer
in the central core region of the tube.   The influence of molecular
viscosity and thermal conductivity will accordingly become important
in the near-wall regions.   In other words a boundary layer flow might
be expected in a relatively thin region close to the wall of the tube.
This assumption, that the flow could be divided into two distinct regions, was the basic premise on which the analysis of Mori and
Nakayama (1967) was subsequently founded.

If the cross stream velocity components are non-dimensionalised in
a similar manner to that used for the axial velocity by Morris (1965b)
we may write

$$U = \frac{ua}{\upsilon} \, , \quad V = \frac{va}{\upsilon} \tag{3.51}$$

The axial momentum equation (3.18) may now be applied in a truncated form to the core region of the flow by omitting the viscous terms to give

$$U_c \frac{\partial W_c}{\partial R} + \frac{V_c}{R} \frac{\partial W_c}{\partial \theta} = \lambda \tag{3.52}$$

where λ is a non-dimensional constant related to the axial pressure gradient by

$$\lambda = - \frac{a^2}{\rho \upsilon^2} \frac{\partial P}{\partial z} \tag{3.53}$$

and c is a sub-script to denote the core flow region.

Similarly if the energy equation (3.19) is evaluated with similar restrictions in the core region it reduces to

$$U_c \frac{\partial \eta_c}{\partial R} + \frac{V_c}{R} \frac{\partial \eta_c}{\partial \theta} = \frac{W_c}{Pr} \tag{3.54}$$

Mori and Nakayama (1967) assumed that the outward cross stream flow has a uniform velocity, D, an assumption based on experimental measurements they had made on other flow systems which involved cross stream secondary flows. This assumption subsequently permitted the following mathematical structures to be postulated for the velocity and temperature distribution in the core region.

$$U_c = D \cos \theta \tag{3.55}$$

$$V_c = D \sin \theta \tag{3.56}$$

$$W_c = A + \frac{\lambda R}{D} \cos \theta \tag{3.57}$$

$$\eta_c = B + \frac{\lambda R^2}{2D^2 Pr} \cos^2 \theta + \frac{AR}{DPr} \cos \theta \tag{3.58}$$

where A and B are constants to be determined later.

It was assumed that the thickness, δ, of the hydrodynamic boundary layer adjacent to the wall could be taken as constant at all angular locations over the duct surface. For convenience, inside the boundary layer radial locations are specified in terms of the corresponding distance, y, measured normally from the wall. Thus

$$y = a - r \tag{3.59}$$

or, in non-dimensional form

$$Y = \frac{y}{a} = 1 - R \tag{3.60}$$

At the outer edge of the hydrodynamic boundary layer the axial velocity must be matched with the core region value. Further the customary zero slip condition must apply at the wall. Mathematically therefore the axial velocity profile inside the boundary layer must satisfy the constraints

$$\left.\begin{array}{l} W = W_{c\Delta} \text{ at } Y = \Delta \\[2ex] \dfrac{\partial W}{\partial Y} = -\dfrac{\partial W_c}{\partial R} \text{ at } Y = \Delta \\[2ex] W = 0 \text{ at } Y = 0 \end{array}\right\} \qquad (3.61)$$

where Δ is the non-dimensional boundary layer thickness given by

$$\Delta = \frac{\delta}{a} \qquad (3.62)$$

and $W_{c\Delta}$ is the velocity of the core flow at $Y = \Delta$. To satisfy these constraints Mori and Nakayama (1967) assumed that the axial velocity distribution in the boundary layer could be expressed as

$$W = W_{c\Delta}\left(2\left[\frac{Y}{\Delta}\right] - \left[\frac{Y}{\Delta}\right]^2\right) + \frac{\Delta\lambda}{D}\left(\left[\frac{Y}{\Delta}\right] - \left[\frac{Y}{\Delta}\right]^2\right)\cos\theta \qquad (3.63)$$

Depending on the Prandtl number of the fluid the thickness of the thermal boundary layer, δ_T, may be greater or less than that of the hydrodynamic counterpart. If the non-dimensional thickness of the thermal boundary layer, Δ_t, is defined in a similar way to that given in equation (3.62) then for $\Delta_T \ll \Delta$ the temperature distribution in the boundary layer must satisfy the constraints

$$\left.\begin{array}{l} \eta = \eta_{c\Delta} \text{ at } Y = \Delta \\[2ex] \dfrac{\partial \eta}{\partial Y} = -\dfrac{\partial \eta_c}{\partial R} \text{ at } Y = \Delta \\[2ex] \eta = 0 \text{ at } Y = 0 \end{array}\right\} \qquad (3.64)$$

where $\eta_{c\Delta}$ is the core flow temperature at the outer edge of the hydrodynamic boundary layer. The constraints of equation (3.64) may be satisfied if, in the boundary layer

$$\eta = \eta_{c\Delta}\left(\frac{2}{\Delta_T}\left[\left[\frac{Y}{\Delta}\right] - 2\left[\frac{Y}{\Delta}\right]^2 + \left[\frac{Y}{\Delta}\right]^3\right] + 3\left[\frac{Y}{\Delta}\right]^2 - 2\left[\frac{Y}{\Delta}\right]^3\right) +$$

$$+ \Delta\left[\frac{A}{D} \cos \theta + \frac{\lambda}{D^2} (1 - \Delta)\right]\left(\left[\frac{Y}{\Delta}\right]^2 - \left[\frac{Y}{\Delta}\right]^3\right) \tag{3.65}$$

Likewise if $\Delta_T \geqslant \Delta$, then

$$\left.\begin{array}{l} \eta = \eta_{c\Delta T} \text{ at } Y = \Delta_T \\[2em] \frac{\partial \eta}{\partial Y} = - \frac{\partial \eta_c}{\partial R} \text{ at } Y = \Delta_T \end{array}\right\} \tag{3.66}$$

which, in turn, requires the boundary layer temperature field to have the structure

$$\eta = \eta_{c\Delta T}\left(2\left[\frac{Y}{\Delta_T}\right] - \left[\frac{Y}{\Delta_T}\right]^2\right) + \Delta_T\left(\frac{\lambda}{D^2} (1 - \Delta_T) + \frac{A}{D} \cos \theta\right) \times$$

$$\times \left(\left[\frac{Y}{\Delta_T}\right]^2 - \left[\frac{Y}{\Delta_T}\right]^3\right) \tag{3.67}$$

Let us now consider the variation of the tangential velocity component within the boundary layer. Figure 3.11 illustrates the nature of the secondary flow distribution considered by Mori and Nakayama (1967). Because there is no net flow across the line OA shown in figure 3.11, continuity dictates that

$$\int_0^\Delta VdY = D(1 - \Delta) \sin \theta \tag{3.68}$$

The tangential velocity distribution in the boundary layer region must also satisfy the boundary conditions

$$\left.\begin{array}{l} V = 0 \text{ at } Y = 0 \\[1.5em] V = V_{c\Delta} \text{ at } Y = \Delta \\[1.5em] \frac{\partial V}{\partial Y} = 0 \text{ at } Y = \Delta \end{array}\right\} \tag{3.69}$$

where $V_{c\Delta}$ is the tangential velocity at the interface between the core and boundary layer flows.

Equations (3.68) and (3.69) may be satisfied if the tangential velocity distribution in the boundary layer is

$$V = - D\left(\left[6 - \frac{12}{\Delta}\right]\left[\frac{Y}{\Delta}\right] + \left[\frac{24}{\Delta} - 9\right]\left[\frac{Y}{\Delta}\right]^2 + \left[4 - \frac{12}{\Delta}\right]\left[\frac{Y}{\Delta}\right]^3\right) \sin \theta \tag{3.70}$$

58

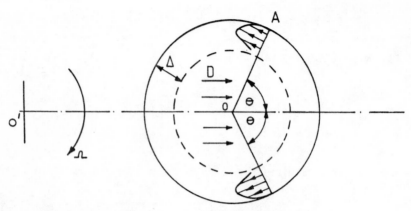

FIG. 3.11   SECONDARY FLOW DISTRIBUTIONS ASSUMED
BY MORI AND NAKAYAMA (1967).

The assumptions made so far have resulted in mathematical forms
being postulated for the velocity and temperature fields in the core
and boundary layer regions. These profiles have implied constants A,
B and λ which are linked to the thickness of the hydrodynamic and
thermal boundary layers and also the uniform cross stream velocity, D,
in the core. Further links between these terms will now be established. The through flow Reynolds number, Re, is related to the non-
dimensional axial velocity profile by

$$Re = \frac{2}{\pi} \int_0^{2\pi} \int_0^1 WRdRd\theta =$$

$$= \frac{2}{\pi}\left\{ \int_0^{2\pi} \int_0^{1-\Delta} WRdRd\theta + \int_0^{2\pi} \int_0^{\Delta} W(1 - Y)dYd\theta \right\} \qquad (3.71)$$

Equation (3.71) may be integrated after the substitution of the core
and boundary layer axial velocity profiles given by equations (3.57)
and (3.63) respectively. Further algebraic manipulation subsequently
permits the following equation for the constant A to be developed

$$A = \frac{Re}{2}\left[\frac{1}{1 - \frac{2}{3}\Delta + \frac{1}{6}\Delta^2} \right] \qquad (3.72)$$

The pressure drop along the tube is in balance with the resistance
offered to flow by the integrated wall shear since developed flow is
considered. This force balance for an elemental tube implies that

$$\frac{\partial p}{\partial z} = \frac{2\mu}{\pi d} \int_0^{2\pi} \left[\frac{\partial w}{\partial y}\right]_{y=0} d\theta \qquad (3.73)$$

which, in non-dimensional form, becomes

$$\lambda = \frac{1}{\pi} \int_0^{2\pi} \left[\frac{\partial W}{\partial Y}\right]_{Y=0} d\theta \qquad (3.74)$$

Substitution of equation (3.63) into equation (3.74) enables integration to be effected. This gives, using equation (3.72)

$$\lambda = \frac{4A}{\Delta} = \frac{2Re}{\Delta}\left[\frac{1}{1 - \frac{2}{3}\Delta + \frac{1}{6}\Delta^2}\right] \qquad (3.75)$$

An energy balance applied in overall terms to the tube links the wall to fluid heat transfer to the enthalpy increase of the fluid. Thus

$$-\frac{kd}{2} \int_0^{2\pi} \left[\frac{\partial T}{\partial r}\right]_{r=a} d\theta = \rho c_p \int_0^{2\pi} \int_0^a w \frac{\partial T}{\partial z} r \, drd\theta \qquad (3.76)$$

Introducing the non-dimensionalisation scheme being used for this problem permits equation (3.76) to be re-expressed as

$$\int_0^{2\pi} \left[\frac{\partial \eta}{\partial R}\right]_{R=1} d\theta = \frac{\pi}{2} Re \qquad (3.77)$$

Depending on the relative thickness of the hydrodynamic and thermal boundary layers, equations (3.65) and (3.67) permit the integral on the left hand side of equation (3.77) to be determined. Noting that Δ and Δ_T are small in comparison with unity permits some simplification of the resulting algebra so that, for both cases, the following expression for B emerges

$$B = \frac{\Delta_T Re}{8} - \frac{Re}{2 D^2 \Delta Pr} \qquad (3.78)$$

The velocity and temperature fields now involve three unknowns in the form of D, Δ and Δ_T respectively. Mori and Nakayama (1967) determined these unknowns by considering the integral form of the axial and tangential momentum equations in the boundary layer together with the appropriate energy equation.

The application of boundary layer approximations to the axial momentum equation (3.18) gives

$$- U\frac{\partial W}{\partial Y} + V\frac{\partial W}{\partial \theta} = \lambda + \frac{\partial^2 W}{\partial Y^2} \tag{3.79}$$

Integration of equation (3.79) across the boundary layer enables the momentum integral equation for the axial direction to be generated as

$$\frac{\partial W}{\partial Y}\bigg]_{Y=0} = W_{c\Delta}\frac{\partial}{\partial \theta}\int_0^\Delta VdY - \frac{\partial}{\partial \theta}\int_0^\Delta VWdY + \lambda\Delta \tag{3.80}$$

This equation may be integrated using equations (3.63) and (3.70) for the axial velocity and tangential velocity profiles respectively. Thus

$$\frac{\partial W}{\partial Y}\bigg]_{Y=0} = E + F \cos\theta \tag{3.81}$$

where

$$E = \left\{\left[\frac{2}{5} - \frac{13}{15}\Delta\right]\cos^2\theta + \left[\frac{3}{5} - \frac{17}{15}\right]\sin^2\theta + \Delta\right\}\lambda \tag{3.83}$$

and

$$F = \frac{D}{4}\left(\frac{2}{5} - \frac{4}{15}\Delta\right)\lambda \tag{3.83}$$

An examination in detail of the variation of E in equation (3.82) with the angular location permitted Mori and Nakayama (1967) to use a mean value given by

$$E = \frac{\lambda}{2} \tag{3.84}$$

This result was also used in their work on flow in curved tubes (see Mori and Nakayama (1965)) and permits equation (3.81) to be simplified to

$$\frac{\partial W}{\partial Y}\bigg]_{Y=0} = \frac{\lambda}{2} + F \cos\theta \tag{3.85}$$

Equation (3.63) may also be used to derive an expression for $\dfrac{\partial W}{\partial Y}\bigg]_{Y=0}$ as

$$\frac{\partial W}{\partial Y}\bigg]_{Y=0} = \frac{\lambda}{2} + \left[\frac{2}{\Delta} - 1\right]\frac{\lambda}{D}\cos\theta \tag{3.86}$$

Comparison of equations (3.85) and (3.86) using equation (3.83) gives

$$(1 - \frac{2}{3}\Delta)D^2\Delta^2 + 10\Delta = 20 \tag{3.87}$$

Mori and Nakayama (1967) simplified this equation noting that $\Delta \ll 1$ to give

$$\Delta = \frac{\sqrt{20}}{D} \tag{3.88}$$

In a similar way the energy equation (3.19) can be truncated to boundary layer form to yield

$$Pr\left[U\frac{\partial \eta}{\partial Y} - V\frac{\partial \eta}{\partial \theta}\right] + \frac{\partial^2 \eta}{\partial Y^2} + W = 0 \tag{3.89}$$

For the case where $\Delta_T \leqslant \Delta$ equation (3.89) may be integrated across the hydrodynamic boundary layer to give

$$\frac{\partial \eta}{\partial Y}\bigg]_{Y=0} = \left[\eta_{c\Delta} \frac{\partial}{\partial \theta} \int_0^\Delta VdY - \frac{\partial}{\partial \theta} \int_0^\Delta \eta VdY\right] Pr + \int_0^\Delta WdY \tag{3.90}$$

Adopting a procedure similar to that outlined for the treatment of the momentum integral equation in the axial direction Mori and Nakayama (1967) proceeded to show that for $Pr \geqslant 1$

$$\frac{\Delta_T}{\Delta} = \frac{2}{11}\left[1 + \sqrt{1 + \frac{77}{4}\frac{1}{Pr^2}}\right] \tag{3.91}$$

Similarly for the case $\Delta_T \geqslant \Delta$ or $Pr \leqslant 1$ the same method gives

$$\frac{\Delta_T}{\Delta} = \frac{1}{5}\left[2 + \sqrt{\frac{10}{Pr^2} - 1}\right] \tag{3.92}$$

The variation of the relative thickness of the thermal and hydrodynamic boundary layers implied by equations (3.91) and (3.92) is shown in figure 3.12.

In order to finally relate the boundary layer thickness to the strength of the cross stream secondary flow generated, that is to the velocity term D, the authors examined the equations of motion for the r and θ directions. In this respect Mori and Nakayama (1967) took the original equations of Morris (1965b) but did not include the buoyancy effect in the Coriolis terms. This is permissible as a result of the discussion presented in Chapter 2.

Some confusion arises in the original paper of Mori and Nakayama (1967) at this stage owing to the treatment of Coriolis force. If the temperature-density variation is not included in the Coriolis term then, as shown in section 3.2, this effect does not contribute to the

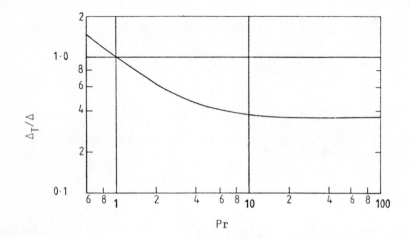

FIG. 3.12   EFFECT OF PRANDTL NUMBER ON THE RELATIVE
THICKNESS OF THE THERMAL AND HYDRODYNAMIC
BOUNDARY LAYERS. MORI AND NAKAYAMA (1967).

creation of secondary flow for the present problem. Mori and Nakayama (1967) still include this effect in their analysis however by
assuming that the core flow will be deflected slightly in the angular
direction. For clarity of presentation this effect is not included
here but will be discussed later in the chapter in more detail.

Mori and Nakayama (1967) used the following truncated version of
equations (3.16) and (3.17) for momentum conservation in the radial
and tangential directions

$$-V^2 \; - \; 4J_a V = \frac{\partial P}{\partial Y} + Ra_\tau \eta \cos \theta \tag{3.93}$$

$$-U\frac{\partial V}{\partial Y} + V\frac{\partial V}{\partial \theta} + 4J_a U = \; - \; \frac{\partial P}{\partial \theta} + \frac{\partial^2 V}{\partial Y^2} - Ra_\tau \eta \sin \theta \tag{3.94}$$

where the dimensionless pressure, P, in the boundary layer is defined
in accordance with equation (3.53) and $J_a = \dfrac{\Omega a^2}{2\nu}$.

The tangential momentum integral equation may be obtained from equation (3.94) by integration across the boundary layer and assuming that
the pressure term is well approximated by that existing at the edge of
the boundary layer which is itself determined from a core-type application of equation (3.17). Thus

$$\left.\frac{\partial V}{\partial Y}\right]_{Y=0} = V_c \frac{\partial}{\partial \theta} \int_0^\Delta V dy - \frac{\partial}{\partial \theta} \int_0^\Delta V^2 dy - 2J_a \int_0^\Delta U dy \; -$$

$$-\Delta \frac{\partial P_{c\Delta}}{\partial \theta} - Ra_\tau \int_0^\Delta \eta \sin \theta \, dY \tag{3.95}$$

By making a number of simplifying assumptions based on their earlier treatment of flow in curved tubes Mori and Nakayama (1967) evaluated equation (3.95) to yield the following relationship between the boundary layer thicknesses and the assumed cross stream flow.
For $Pr \geqslant 1$

$$D = 0.930 \left[3 \frac{\Delta_T}{\Delta} - 1 \right]^{\frac{1}{5}} (Re \ Ra_\tau)^{\frac{1}{5}} \tag{3.96}$$

For $Pr \leqslant 1$

$$D = 1.330 \left[\frac{\Delta_T}{\Delta} - 1 + \frac{\Delta}{3\Delta_T} \right] (Re \ Ra_\tau)^{\frac{1}{5}} \tag{3.97}$$

The variation of friction factor and Nusselt number may now be determined as follows. Making use of the definition of Blasius friction factor as given by equation (3.37) it follows that, in terms of the present nomenclature

$$\frac{C_{fR}}{C_{fo}} = \frac{\lambda}{4 \, Re} \tag{3.98}$$

The pressure gradient term, λ, may be calculated using equations (3.75) and (3.88) which for small values of the boundary layer thickness permits the resistance characteristics to be calculated from

$$\frac{C_{fR}}{C_{fo}} = \frac{D}{2\sqrt{20}} \tag{3.99}$$

The cross stream secondary flow term, D, is determined easily from equations (3.96) or (3.97) as appropriate so that
for $Pr \geqslant 1$

$$\frac{C_{fR}}{C_{fo}} = 0.104 \left[3 \frac{\Delta_T}{\Delta} - 1 \right]^{\frac{1}{5}} (Re \ Ra_\tau)^{\frac{1}{5}} \tag{3.100}$$

with $\frac{\Delta_T}{\Delta}$ calculated from equation (3.91)
and for $Pr \leqslant 1$

$$\frac{C_{fR}}{C_{fo}} = 0.149 \left[\frac{\Delta_T}{\Delta} - 1 + \frac{1}{3} \frac{\Delta}{\Delta_T} \right]^{\frac{1}{5}} (Re \ Ra_\tau)^{\frac{1}{5}} \tag{3.101}$$

64

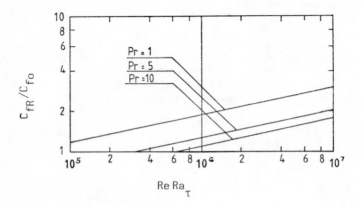

FIG. 3.13   INFLUENCE OF ROTATION ON FLOW RESISTANCE
MORI AND NAKAYAMA (1967).

with $\dfrac{\Delta_T}{\Delta}$ calculated from equation (3.92).

Typical variations of the resistance coefficients are presented in
figure 3.13 for a range of Prandtl number values. Owing to the nature
of the assumptions made these friction factor results do not appear to
be valid at relatively lower values of the Re Ra$_\tau$ product, the actual
range being dependent on the Prandtl number. This is based on the
fact that the ratio $\dfrac{C_{fR}}{C_{fo}}$ is not expected to fall below unity.

The influence of rotation on heat transfer was determined by evaluating the Nusselt number using the definition presented in equation
(3.45). In this respect Mori and Nakayama (1967) calculated the bulk
Nusselt number, $Nu_{\infty,b}$ using the difference between the wall temperature
and the bulk fluid temperature as the motivating potential for heat
transfer. To actually evaluate the bulk temperature these authors
assumed that the contribution of the boundary layer temperature distribution to the bulk value was small. Accordingly the integration
of equation (3.49) for bulk temperature was carried out over the core
region only. This gave

$$(T_w - T_b) = \left[\frac{\Delta_T \, Re \, Pr}{16} + \frac{Re}{8 \, D^2 \Delta} \right] \tau d \qquad (3.102)$$

Noting that at zero rotation speed the Nusselt number, Nu_o, is obtained
from the classic Nusselt solution of this problem as

$$Nu_o = \frac{48}{11} \qquad (3.103)$$

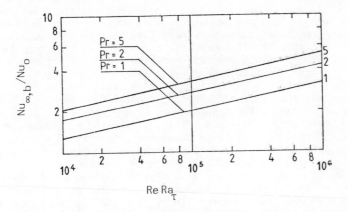

FIG. 3.14   INFLUENCE OF ROTATION ON HEAT TRANSFER.
MORI AND NAKAYAMA (1967).

we get using the previously derived results

$$\frac{Nu_{\infty,b}}{Nu_o} = \frac{11}{12\sqrt{20}} \frac{D\Delta}{\Delta_T} \left[\frac{1}{1 + \left[\frac{\Delta}{10\Delta_T Pr}\right]}\right] \tag{3.104}$$

Subsequent evaluation of equation (3.104) using the expression for Δ, Δ_T and D as appropriate to the Prandtl number range required permits the final result for heat transfer in this rotating tube configuration to be determined.
Thus for Pr $\geqslant$ 1

$$\frac{Nu_{\infty,b}}{Nu_o} = \frac{0.191\Delta}{\Delta_T} \left[\frac{3\Delta_T}{\Delta} - 1\right]^{\frac{1}{5}} (Re\ Ra_\tau)^{\frac{1}{5}} \tag{3.105}$$

with $\frac{\Delta_T}{\Delta}$ calculated using equation (3.91)
and for Pr $\leqslant$ 1

$$\frac{Nu_{\infty,b}}{Nu_o} = \frac{0.273\Delta}{\Delta_T} \left[\frac{\Delta_T}{\Delta} - 1 + \frac{\Delta}{3\Delta_T}\right]^{\frac{1}{5}} (Re\ Ra_\tau)^{\frac{1}{5}} \tag{3.106}$$

with $\frac{\Delta_T}{\Delta}$ calculated using equation (3.92).

The typical effect of rotation on heat transfer resulting from the analysis of Mori and Nakayama (1967) is shown in figure 3.14.   A detailed discussion on these results will follow at the end of this section of the chapter.
Both the analyses described above have their individual limitations.

We have already discussed the fact that the solution method used by Morris (1965b) is only applicable to low heating rates and rotational speeds. The work of Mori and Nakayama (1967) on the other hand is valid provided the cross stream secondary flow is very strong. It is not easy to illustrate clearly when this assumption is true and the way, for example, that fluid Prandtl number influences the range of applicability is difficult to assess. For these reasons Woods and Morris (1974) attempted to resolve the problem using a numerical procedure to solve the basic differential equations of mass, momentum and energy conservation without making any assumptions concerning the nature of the secondary flow field. This, it was argued, would produce a single solution procedure to cover the entire range of rotational speeds, heating rates, Prandtl numbers, etc., within the basic assumption of a uniformly heated and developed laminar flow.

Woods and Morris (1974) used the same basic equations developed originally by Morris (1965b) but instead of using the single biharmonic equation (3.25) involving the stream function for the conservation of radial and tangential momentum elected to use an additional vorticity equation. Thus the equations used were

$$\nabla^2 \Psi = - \xi \tag{3.107}$$

$$\nabla^2 \xi + \frac{1}{R} \frac{\partial(\Psi,\xi)}{\partial(R,\theta)} - Ra_\tau \left[\frac{1}{R} \frac{\partial \eta}{\partial \theta} \cos \theta + \frac{\partial \eta}{\partial R} \sin \theta \right] -$$

$$- \frac{Ra_\tau}{\epsilon_a} \frac{\partial \eta}{\partial \theta} + \frac{Ra_\tau}{J_a \epsilon_a} \frac{1}{R} \frac{\partial(\Psi,\eta)}{\partial(R,\theta)} = 0 \tag{3.108}$$

$$\nabla^2 W + \frac{1}{R} \frac{\partial(\Psi,W)}{\partial(R,\theta)} + 4 Re_p = 0 \tag{3.109}$$

$$\nabla^2 \eta + \frac{Pr}{R} \frac{\partial(\Psi,\eta)}{\partial(R,\theta)} + W = 0 \tag{3.110}$$

Equations (3.107) through (3.110) may be all compared to a general equation having the structure

$$a_\phi \frac{\partial(\phi \ \Psi)}{\partial(R,\theta)} - R\nabla^2\phi + Rd_\phi = 0 \tag{3.111}$$

where, in turn, ϕ stands for the stream function, Ψ, the vorticity, ξ, the axial velocity, W, and the temperature function, η. The expression for the coefficients a_ϕ and d_ϕ respectively are given in Table 3.1.

Gosman et al (1968) derived a computational method for solving simultaneous sets of equations having the general structure of equation (3.111). The application of this method to the present problem will now be briefly described.

Consider a grid network to cover the cross section of the tube so

Φ	a_Φ	d_Φ
Ψ	0	$-\xi$
ξ	1	$Ra_\tau\left[\dfrac{1}{R}\dfrac{\partial\eta}{\partial\theta}\cos\theta + \dfrac{\partial\eta}{\partial R}\sin\theta\right]$ $+\dfrac{Ra_\tau}{\varepsilon_a}\dfrac{\partial\eta}{\partial\theta} - \dfrac{Ra_\tau}{J_a\varepsilon_a}\dfrac{1}{R}\dfrac{\partial(\Psi,\eta)}{\partial(R,\theta)}$
W	1	$-4\,Re_p$
η	Pr	$-W$

TABLE 3.1   IMPLIED COEFFICIENTS FOR EQUATION (3.111).

that there are N radial spaces and M tangential spaces. Grid points in the immediate vicinity of a typical mode, P, are represented by the compass points N,S,W,E, etc., and similar locations midway between grid points by n,s,w,e, etc., as shown in figure 3.15.

Finite difference approximation of equation (3.111) is achieved by integrating the equation over the control rectangle ne - se - sw - nw which surrounds a typical node, P, making suitable assumptions for the variation of the generalised variable, Φ, between the nodes. Precise details of the assumptions made for the Φ-variation may be found in the original work of Gosman et al (1968).

The resulting finite difference equation which links the Φ-value at a particular node to its neighbours has the form

$$\Phi_P = C_E\,\Phi_E + C_W\,\Phi_W + C_N\,\Phi_N + C_S\,\Phi_S + D_P \qquad (3.112)$$

where the subscripted values of Φ refer to the corresponding nodes in the vicinity of P and C_E, C_W, C_N, C_S and D are coefficients which result from the approximating finite difference representation. The mathematical expressions for the coefficients given in equation (3.112) require a considerable amount of algebraic labour for their derivation. Readers are referred to Woods (1975) for details of the manipulations required in their formulation. However in broad terms they have the structures outlined below.

68

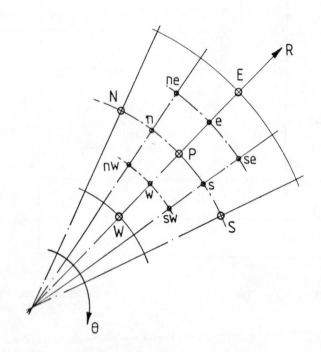

FIG. 3.15   GRID NOTATION FOR FINITE DIFFERENCE
SCHEME WOODS AND MORRIS (1974).

$$C_E = \frac{A_E + B_E}{\Sigma AB}$$

$$C_W = \frac{A_W + B_W}{\Sigma AB}$$

$$C_N = \frac{A_N + B_N}{\Sigma AB}$$

(3.113)

$$C_S = \frac{A_S + B_S}{\Sigma AB}$$

where

$$\Sigma AB = (A_E + A_W + A_N + A_S) + (B_E + B_W + B_N + B_S)$$

(3.114)

and

$$A_E = \frac{a_\phi}{8}\left(A_1 + \left|A_1\right|\right)$$

$$A_W = \frac{a_\phi}{8}\left(A_2 + \left|A_2\right|\right)$$

$$A_N = \frac{a_\phi}{8}\left(A_3 + \left|A_3\right|\right)$$

$$A_S = \frac{a_\phi}{8}\left(A_4 + \left|A_4\right|\right)$$

(3.115)

with

$$A_1 = \Psi_{SE} + \Psi_S - \Psi_{NE} - \Psi_N$$

$$A_2 = \Psi_{NW} + \Psi_N - \Psi_{SW} - \Psi_S$$

$$A_3 = \Psi_{NE} + \Psi_E - \Psi_{NW} - \Psi_W$$

$$A_4 = \Psi_{SW} + \Psi_W - \Psi_{SE} - \Psi_E$$

(3.116)

also

$$B_E = \frac{(R_P + R_E)(\theta_N - \theta_S)}{4 \quad (R_E - R_P}$$

$$B_W = \frac{(R_P + R_W)(\theta_N - \theta_S)}{4 \quad (R_P - R_W)}$$

$$B_N = \frac{(R_E - R_W)}{2R_P(\theta_N - \theta_P)}$$

$$B_S = \frac{(R_E - R_W)}{2R_P(\theta_P - \theta_S)}$$

(3.117)

The remaining coefficient D_P is given by

$$D_p = - \frac{d_{\phi p} R_p (\theta_N - \theta_S)(R_E - R_W)}{4 \ \Sigma AB} \qquad (3.118)$$

where d_ϕ is evaluated from Table 3.1 using first order central difference approximations for all gradients required.

At the centre point of the cylindrical grid system used the succes-
sive substitution formula given by equation (3.112) cannot be used.
At this location the equations exhibit a singularity. To overcome
this Woods and Morris (1974) re-expressed the equations at the centre
point in Cartesian coordinates and hence derived a special form of
equation (3.112) appropriate to the central node.

The boundary conditions for stream function, axial velocity and temperature at the wall of the duct are Ψ = constant, W = 0 and η = 0
and it is convenient to take Ψ = 0 for the stream function. The vorticity boundary condition causes problems however. To overcome this
Morris and Woods (1974) assumed that in the near wall region the axial
gradients of all dependent variables were small in relation to cross
stream gradients. This permitted the full conservation equations to
be truncated to a set of ordinary differential equations valid in the
vicinity of the wall. These equations were

$$\left.\begin{array}{l} \dfrac{d}{dR}\left[R \ \dfrac{d\Psi}{dR} \right] = - R\xi \\[3mm] \dfrac{d}{dR}\left[R \ \dfrac{d\xi}{dR} \right] = Ra_\tau \ \sin\theta \ R \ \dfrac{d\eta}{dR} \\[3mm] \dfrac{d}{dR}\left[R \ \dfrac{d\eta}{dR} \right] = - RW \\[3mm] \dfrac{d}{dR}\left[R \ \dfrac{dW}{dR} \right] = - 4 \ R \ Re_p \end{array}\right\} \qquad (3.119)$$

This set of equations could be integrated sequentially to give the
vorticity values at the wall in terms of vorticity, stream function,
axial velocity and temperature at nodes immediately adjacent to the
wall for any value of angular location. If the subscript "wall" is
used for a wall-value of vorticity and "o" for a neighbouring point in
the radial direction then integration of equation (3.119) yields

$$\xi_{wall} = E_1 \ \xi_o + E_2 \ \Psi_o + Ra_\tau \ \sin\theta \ E_3 \ \eta_o + E_4 \ W_o + E_5 \ Re_p \qquad (3.120)$$

where the coefficients E_1 through E_5 are given by

$$E_1 = F_4H_2/(F_4H_2 - F_5)$$

$$E_2 = 1/(F_4H_2 - F_5)$$

$$E_3 = [H_2F_3 + F_4H_2H_7]/(F_4H_2 - F_5)$$

$$E_4 = [F_2H_2 + F_3H_4H_2 + F_4H_6H_2 + F_4H_7H_4H_2]/(F_4H_2 - F_5)$$

$$E_5 = [F_1 + F_2H_1 + F_3H_3 + F_3H_4H_1 + F_4H_5 + F_4H_6H_1 +$$
$$+ F_4H_7H_3 + F_4H_7H_4H_1]/(F_4H_2 - F_5)$$

(3.121)

where

$$F_1 = [36 R_o^7 - 392 R_o^5 + 2009 R_o^2 - 2310 \ln R_o -$$
$$- 1653]/176400$$

$$F_2 = [R_o^5(188 - 120 \ln R_o) - 875 R_o^2 = 930 \ln R_o +$$
$$+ 687]/54000$$

(3.122)

$$F_3 = [4 R_o^3 - 9 R_o^2 + 6 \ln R_o + 5]/36$$

$$F_4 = [R_o^2(\ln R_o - 1) + \ln R_o + 1]/4$$

$$F_5 = [R_o^2 - 2 \ln R_o - 1]/4$$

and

$$H_1 = [R_o^2 - 1]/\ln R_o$$

$$H_2 = 1/\ln R_o$$

$$H_3 = [4 R_o^2 - R_o^4 - 3]/16 \ln R_o$$

$$H_4 = [R_o^2(\ln R_o - 1) + 1]/4 \ln R_o$$

(3.123)

$$H_5 = [50 \ R_o^{\ 3} - 9 \ R_o^{\ 5} - 41]/900 \ \ell n \ R_o$$

$$H_6 = [R_o^{\ 3}(6 \ \ell n \ R_o - 7) + 7] \ /105 \ \ell n \ R_o$$

$$H_7 = [1 - R_o]/\ell n \ R_o$$

Note that R_o is the radius level corresponding to nodes immediately next to the wall location and is determined once the grid spacing has been selected.

The finite difference equations at each internal node including the centre point together with those at the boundary form a set of non-linear algebraic equations. These equations were solved using a Gauss-Seidel iterative procedure in which the variable coefficients required were up-dated after each iterative cycle. The conditions for convergence of this type of equation set have not been established formally. However, adopting the method of Gosman et al (1968), the convergence criteria for a linear set of equations was used as a guide. Provided the variation in the coefficients responsible for the non-linearity was not too great experience demonstrated that this procedure was suitable in most instances. Once the solution had converged the friction factor and Nusselt number were determined by numerical integration of the appropriate defining equation.

The theoretical model seemingly requires that Ra_τ, Re_p, Pr, J_a and ε_a are independently specified. However Woods (1975), Woods and Morris (1981) have shown that the product $Re_p \ Ra_\tau$ is a naturally emerging group so the independent specification of Ra_τ and Re_p is unnecessary. This reduces the amount of computational labour needed to examine the effect of various parameters influencing the problem. Note, as with other attempts to solve this problem, once a particular running condition has been specified the through flow Reynolds number, Re, emerges from the solution for the axial velocity field. Because this form of Reynolds number is more directly amenable to physical interpretation than its pseudo counterpart Morris and Woods (1974) adopted the product $Re \ Ra_\tau$ with which to present the results of their study.

A variety of sample calculations were performed prior to the production computations in order to gain experience of the numerical procedure and to assess its accuracy. Calculations performed at particular running conditions with different finite-difference grid distributions confirmed that the solution technique had good convergence characteristics. As a result of these exploratory calculations, a grid distribution involving 16 equispaced radial and tangential steps (16 x 16) was deemed to give a suitable compromise between accuracy and computing time.

As mentioned earlier the assumptions made in the analysis are those of the classical Nusselt (1910) problem when the tube is stationary. The known solution in this instance gives a bulk Nusselt number of

48/11 and the numerical procedure agreed with this figure to within
0.1 per cent with a 16 x 16 grid.

It was found that the influence of Coriolis acceleration, characterised by the rotational Reynolds number, J_a, had little effect on
the solution. Because of this insensitivity, J_a was set at 10^9 during
the calculations; this effectively removed the Coriolis term from
the controlling equations.

With air as the convective fluid and a running condition selected
to produce strong buoyancy effects (Re_p Ra_τ = 10^7, ε_a = ∞, J_a = 10^9),
it was found that the values of Nusselt number, friction factor,
centre-line axial velocity and centre-line temperature obtained with
a 16 x 16 grid were within 2 to 4 per cent of a converged solution
obtained with more refined grid systems. The agreement was considerably better for the weaker secondary flows which occurred at lower
values of Re_p Ra_τ and higher values of Prandtl number. The improved
accuracy achieved with higher values of Prandtl number and a specified
grid node distribution may be accounted for by the fact that the reduction in error arising from the associated decreasing strength of
the secondary flow offsets the increase in error which arises from
the confinement of the thermal boundary layer to a region nearer to
the wall.

The relative improvement in heat transfer which results from rotation of the tube is typified by the results shown in figure 3.16.
Here the ratio of Nusselt numbers obtained under rotating and non-rotating conditions is plotted against the product Re Ra_τ for an infinite
eccentricity. This eccentricity was selected because, as shown later,
the solution becomes independent of eccentricity when ε_a > 5.

With air (Pr = 0.73), significant improvement in heat transfer is
apparent. Indeed at Re Ra_τ = 10^6, the heat transfer is increased by
about 200 per cent compared with the non-rotating case. At this value
of Prandtl number there is quite good agreement with the predictions
of Mori and Nakayama (1967), although these authors suggest a steeper
gradient than that of the more-exact numerical solution. Mori and
Nakayama (1967) predict that $Nu_{\infty,b} \propto (Re\ Ra_\tau)^{1/5}$ for Pr = 0.73, whereas the present analysis suggests that $Nu_{\infty,b} \propto (Re\ Ra_\tau)^{1/5 \cdot 5}$. Improvements in the heat transfer are also evident for fluids having Prandtl
numbers greater than those associated with gases. For a given value
of Re Ra_τ, the greatest improvements occur at the higher values of
Prandtl number. It is at these higher values of Prandtl number that
the present predictions show serious quantitative discrepancies with
those of Mori and Nakayama (1967), although the same qualitative
trends are evident regarding the influence of Prandtl number.

The variation in resistance to flow, expressed as the ratio of friction factor to its non-rotating value, is shown in figure 3.17. Significant increases in resistance are evident, particularly at low
values of Prandtl number. The predictions of Mori and Nakayama (1967)
for flow resistance are not in good agreement with the present analy-

74

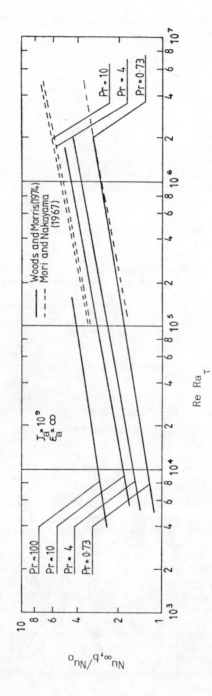

FIG. 3.16   INFLUENCE OF ROTATION ON HEAT TRANSFER WOODS AND MORRIS (1974).

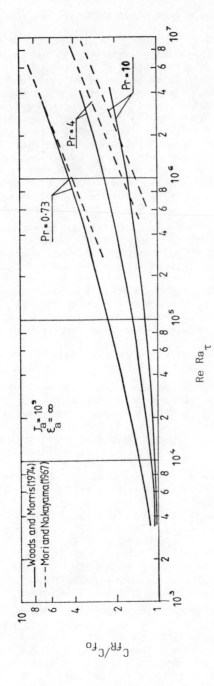

FIG. 3.17   INFLUENCE OF ROTATION ON FLOW RESISTANCE. WOODS AND MORRIS (1974).

sis, although once more the same qualitative influence of Prandtl
number is evident.

The general quantitative disagreement with the data of Mori and
Nakayama (1967) may be explained by reference to the assumptions made
by these authors concerning the nature of the secondary-flow distribution and the associated effect on the axial velocity and temperature
profiles. Mori and Nakayama's integral technique assumes a core region in which there exists a strong outward secondary flow and a thin
inward-flowing boundary layer in the vicinity of the wall which permits the fluid to be eventually re-circulated into the core. Associated with this assumption is an axial velocity distribution comprising
a series of ruled lines in the core region. These assumptions and the
implied results are claimed to be valid for high rotational speeds.
Figures 3.18 and 3.19 show actual stream function, axial velocity and
temperature distributions obtained from the present analysis. Refer-
ring to figure 3.18, in which $Re_p Ra_\tau = 10^3$, the stream function distribution shows that the assumption of a strong core secondary flow
with a thin boundary layer for re-circulation is not true, nor is the
implied assumption of an axial velocity distribution which is linear
in the radial direction. This is true for both Prandtl numbers used
for the illustrative comparisons. In both cases, the cross-stream
buoyancy results in a slight distortion of the axial velocity and temperature profiles from their usual axisymmetric form for pipe flows.
This behaviour is typical of the weak secondary flow induced by the
earth's field in a horizontal pipe (see the perturbation analysis of
Morton (1959)).

Velocity and temperature distributions are shown in figure 3.19 for
the case where $Re_p Ra_\tau = 10^6$. Once more, for both Prandtl numbers
shown, the region of return flow along the walls appears to be far too
thick to be described by boundary layer approximations. Nevertheless,
for fluids having Prandtl numbers near unity, the axial velocity and
temperature distributions in the core region bear some resemblance to
that assumed by Mori and Nakayama (1967) (see the distribution obtained for Pr = 0.73 in figure 3.19). At high Prandtl numbers, the viscosity of the fluid appears to inhibit secondary flow. This is indicated in figure 3.19, where the maximum value of the stream function
for a Prandtl number of 4 is approximately half that obtained with a
Prandtl number of 0.73. The assumption of a core region in which
axial momentum and enthalpy is convected predominantly by the secondary flow is not generally valid. It is for this reason that the heat
transfer predictions of Mori and Nakayama (1967) are in close agreement with the present predictions only when Pr = 0.73. In order for
the technique suggested by Mori and Nakayama (1967) to be used for
fluids of high Prandtl number, it would presumably be necessary to
operate at exceedingly high values of $Re\, Ra_\tau$, under which

circumstances transition to turbulence may have occurred.

The difference in distributions of axial velocity for Prandtl numbers of 0.73 and 4.0 demonstrates clearly the effect of Prandtl number on flow resistance. At lower Prandtl numbers the distortion of
the axial field becomes much more severe. The flattening of the axial
velocity profile results in a far lower flow rate for a given pressure

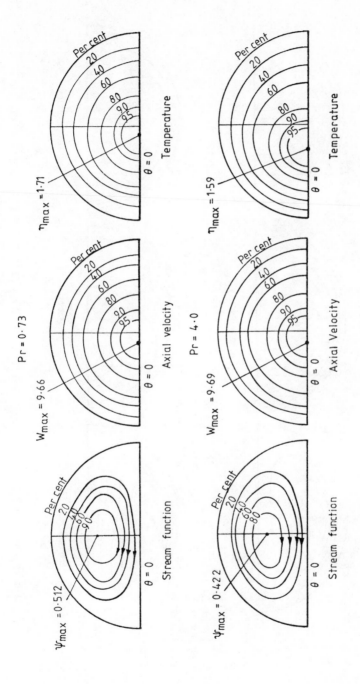

FIG. 3.18   VELOCITY AND TEMPERATURE FIELDS FOR A RELATIVELY WEAK SECONDARY FLOW. WOODS AND MORRIS (1974) (Re_p Ra $_\tau$ = 10^3, ε_a = ∞, J_a = 10^9).

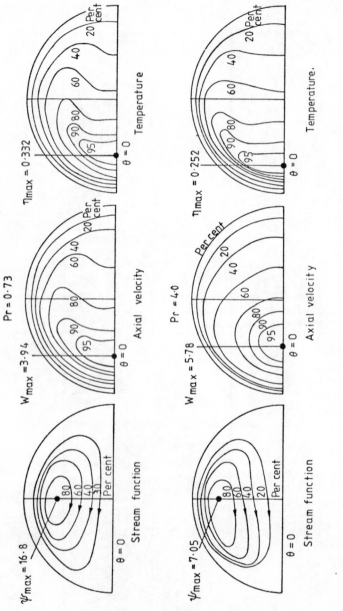

FIG. 3.19   VELOCITY AND TEMPERATURE FIELDS FOR A RELATIVELY STRONG SECONDARY FLOW. WOODS AND MORRIS (1974) (Re_p $Ra_\tau = 10^6$, $\varepsilon_a = \infty$, $J_a = 10^9$).

gradient and thus the increase in friction coefficient. It is interesting to note that Siegworth et al (1969) in their studies of combined forced and gravitation-induced free convection in horizontal
tubes, use dimensional arguments to demonstrate that as the Prandtl
number of the fluid approaches infinity the axial velocity profile
becomes unaffected by the cross-stream buoyancy. This is confirmed
in the present analogous study. For Prandtl numbers of 10^2 and 10^3,
it was found that the friction coefficient was less than 1 per cent
greater than that for the stationary case. To exemplify, the distortion of the axial velocity profile is less severe at the higher of
the two Prandtl numbers shown. This point will be treated in greater
depth later in this section.

Numerical experimentation with the solution programme permitted the
influence of eccentricity and Coriolis acceleration to be examined.
It was found that the eccentricity of the tube, expressed in terms of
the eccentricity parameter ε_a, had little influence on the heat transfer and flow resistance provided that $\varepsilon_a > 5$. In relation to the
problems of cooling electrical machine rotors it is interesting to
note that normally $\varepsilon_a > 50$. For $\varepsilon_a < 5$ there was found a small increase in both Nusselt number and friction factor but the effect is
small enough to ignore in the vast majority of rotor cooling applications.

In Chapter 2 it was shown that to a first approximation the interaction between the Coriolis acceleration and the variable density
could be omitted. Morris (1965b) and Woods and Morris (1974) did however permit such an interaction and it is interesting now to discuss
the Coriolis influence in more detail.

The Coriolis effect is controlled by the rotational Reynolds number,
J_a, and this parameter influences the vorticity equation (3.108) via
a term having the structure

$$\frac{Ra_\tau}{J_a \varepsilon_a} \frac{1}{R} \frac{\partial(\Psi, \eta)}{\partial(R, \theta)}$$

This suggests that Coriolis acceleration will become important at low
values of rotational Reynolds number and eccentricity parameter. This
is a contradictory result in relation to the analysis of Mori and
Nakayama (1967).

Figure 3.20 typifies the influence of Coriolis acceleration on heat
transfer for the case where $Re_p Ra_\tau = 10^6$ and $\varepsilon_a = 1$. For $J_a > 80$
the effect is negligible for all the Prandtl number values shown.
When $J_a < 80$ the Nusselt number is reduced with the attentuation more
marked at the low Prandtl number values. Figure 3.21 shows the corresponding effect on flow resistance. For eccentricity levels typical
of electrical machine rotor applications and the corresponding values
of J_a it is unlikely that Coriolis acceleration will significantly
affect developed laminar heat transfer and flow resistance.

Figure 3.22 shows how the Coriolis acceleration can influence the

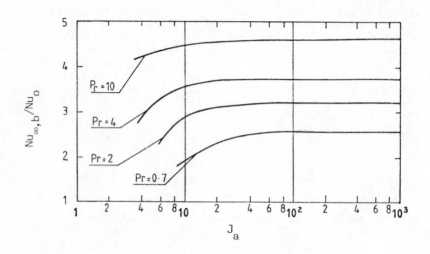

FIG. 3.20   INFLUENCE OF CORIOLIS ACCELERATION ON
HEAT TRANSFER. WOODS (1975).

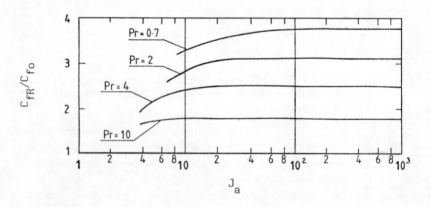

FIG. 3.21   INFLUENCE OF CORIOLIS ACCELERATION ON
FLOW RESISTANCE. WOODS (1975).

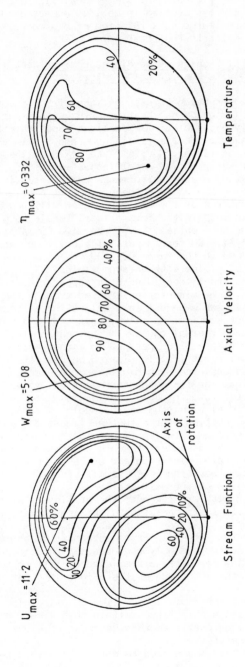

FIG. 3.22   TYPICAL DISTORTION OF SECONDARY FLOW PROFILE DUE TO CORIOLIS ACCELERATION. WOODS (1975) (Re_p Ra_τ = 10^6, ϵ_a = 1, Pr = 2, J_a = 10).

flow and temperature fields. The symmetrical flow patterns in both halves of the tube cross section is now, as expected, destroyed. The distortion is approximated by a rotation of a symmetric solution in the opposite direction to the rotation.

3.3.3 Experimental Studies of Developed Laminar Heat Transfer and Comparison with Theoretical Predictions.

We now turn our attention to a comparison of the theoretical predictions for developed laminar heat transfer with experimental data which is currently available. The earliest reported experimental study was undertaken by Morris (1964) and reported in various forms by Davies and Morris (1965), Morris (1968, 1970). These works which reported data for water and glycerol respectively were originally undertaken as a study of the performance characteristics of a particular form of rotating closed loop thermosyphon. A brief description of the salient features of the apparatus used will now be given.

A tubular closed loop in the form of a rectangle ABCD was constrained to rotate about the centre line of the limb CD, as illustrated in figure 3.23. The limb CD formed part of a built up rotor which could be rotated at speeds in the range 0 - 300 rev/min. The circuit ABCD could be completely filled with various fluids.

The test section itself consisted of a copper rod 25.4 mm external diameter with a 6.35 mm diameter hole bored through the centre. Pyrotenax heating cable was embedded in a helical groove machined on the outside of the test section and silver soldered at a number of positions to ensure good thermal contact. The heating cable was made from Nichrome resistance wire sheathed in a thin-walled tube of stainless steel. Mutual electrical insulation between the resistance wire and the sheath was achieved with highly compressed magnesium oxide powder. The nominal length of the test section was 304.8 mm, giving a length to diameter ratio of 48.

Nickel chrome/nickel aluminium thermocouples were soldered at both ends of the heated portion of the test section and also at a location 24 diameters downstream of the entry station. It was at this location that the heat transfer measurements were evaluated. Both ends of the test section were sealed with threaded end caps which were fitted with a copper/constantan thermocouple to permit measurement of the fluid temperature at the inlet and exit stations.

Thermocouple signals were taken to the stationary measuring equipment via a miniature instrumentation slip ring assembly located at one end of the rotor. External heat loss was reduced by covering the outside of the heater with a layer of refractory cement approximately 6 mm thick. The heater assembly could easily be removed from the apparatus and the flow circuit was completed by fitting the short copper tubes seen at the inlet and exit stations into perspex radial limbs BC and DA.

During operation, the fluid in the limb CD was cooled using mains water flowing inside a coil fitted within this limb. Rotary seals at each end of the rotor permitted this secondary coolant to flow into and out of the apparatus.

The fluid under test was caused to circulate within the closed loop

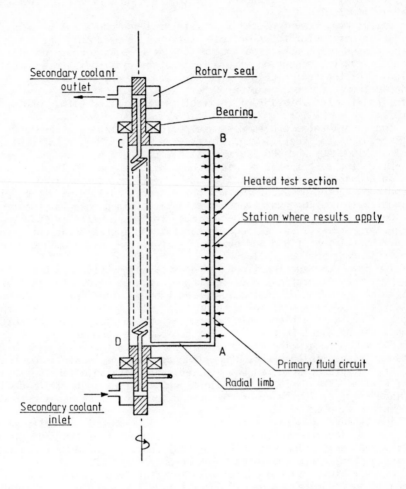

FIG. 3.23   SCHEMATIC LAYOUT OF APPARATUS USED BY
            MORRIS (1964).

owing to the well known thermosyphon effect, the rate of flow being governed by the heat transfer from the test section for a given geometry of flow.

Direct measurement of the flow rate achieved could not be performed in this particular rotating test section. However this parameter was calculated using a heat balance method after making measurements of heater power consumption, fluid temperature rise and external heat loss from the test section. The heat loss at any operating condition was determined from a series of heat loss calibration experiments performed with the interior of the test section filled with granulated cork.

Two experimental programmes were performed using water and 100% glycerol as the test fluids. Tests were conducted at rotational speeds of 50, 100, 200 and 300 rev/min which, for the apparatus used, gave centre-line centripetal accelerations in the range 0 - 15g.

To support their theoretical study, Woods and Morris (1974) designed and constructed an apparatus which was specifically aimed at supplying local and mean heat transfer data with this particular rotating flow geometry. The apparatus was much more sophisticated than that used by Morris (1964) as regards control and instrumentation. Air was used as the test fluid and brief details of the apparatus and experimental programme now follow.

The main constructional details of the test sections used for the experimental investigations are shown in figure 3.24. The test section comprised an electrically-heated tube mounted inside an encapsulating cylinder and fitted with an inlet plenum and exit mixing chamber. The test section could be mounted onto a slave rotor so that its axis was parallel to the axis of rotation, the eccentricity being 0.305 m. A controlled electric motor permitted the rotor assembly to run at speeds in the range 100 - 1500 rev/min steplessly. Two test sections were constructed and studied, each with an effective length of 0.61 m. The first of these test sections (designated A) had a bore diameter of 12.7 mm whereas the second (designated B) had a bore diameter of 6.35 mm.

The coolant was drawn through the test section by a centrifugal fan, the flow circuit comprising a Rotameter flowmeter, inlet and exit rotary seal units plus necessary internal ducting. A schematic arrangement of the entire apparatus is shown in figure 3.25.

Chromel/Alumel thermocouples were mounted along the test section to enable an accurate determination of the wall temperature distribution to be made. Also, a number of thermocouples were mounted in the inlet and exit chambers so permitting the air temperature rise to be measured. All these rotor-mounted thermocouples could be monitored sequentially using a data logger, signals being taken from the rig via silver-silver graphite instrumentation slip-ring assemblies mounted at each end of the rotor shaft.

A separate power slip-ring comprising two phosphor-bronze rings with spring-loaded carbon brushes enabled power to be taken to the heated wire. The rotational speed was measured with a magnetic transducer and electronic timer.

Test section A (bore diameter 12.7 mm) was tested at nominal Reynolds numbers of 1300, 1900 and 2500 and eleven rotational speeds

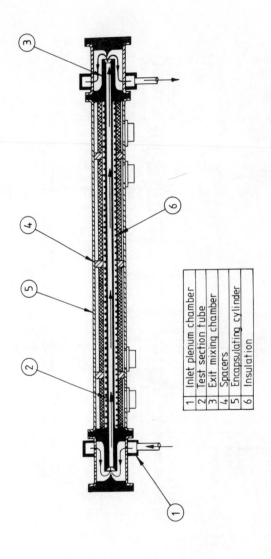

1	Inlet plenum chamber
2	Test section tube
3	Exit mixing chamber
4	Spacers
5	Encapsulating cylinder
6	Insulation

FIG. 3.24   A SCHEMATIC REPRESENTATION OF THE TEST SECTION USED BY WOODS AND MORRIS (1974).

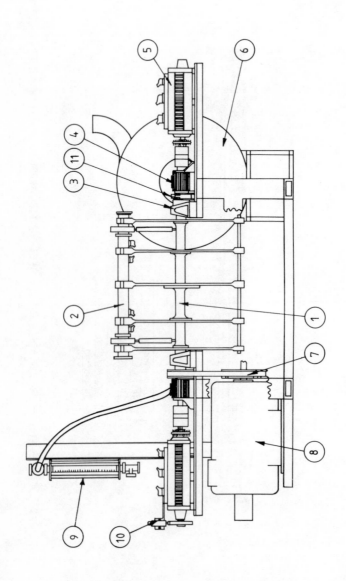

FIG. 3.25   A SCHEMATIC REPRESENTATION OF THE APPARATUS USED BY WOODS AND MORRIS (1974).

equispaced in the range 0 - 1000 rev/min. Test section B (bore diameter 6.35 mm) was tested at Reynolds numbers of 1900 and 2500 and
six rotational speeds equispaced in the range 0 - 1000 rev/min. At
each speed and Reynolds number pair, five different heater power settings were used.

It was found that a thermal setting time of about one hour was necessary for each run, after which all the wall and fluid temperatures
were recorded on punched tape and the readings of rotational speed,
heating rate, air flow rate, and ambient conditions were noted.

To ensure confidence in the test readings, it was necessary to mon-
itor the accuracy of the experiments by checking that the overall heat
balance was adequately within an acceptable error bound. This was
done by comparing the measured exit air bulk temperature against that
calculated using the measured heater power, the air flow rate and a
previously determined set of heat loss calculations. Generally, the
agreement between the measured and the calculated exit bulk temperature was within 10 per cent. This was deemed to be of adequate accuracy considering the nature of the experiment.

Although the energy dissipation rate from the heating element is, to
a good approximation, uniform along the tube length, it does not necessarily follow that the actual axial distribution of heat flux to the
fluid is correspondingly uniform. This is due to the axial variation
in wall temperature and its associated influence on the axial distribution of heat loss. For this reason the bulk temperature variation
along the tube (even with an invariant, constant-pressure specific
heat) may depart from the linear form which occurs with a truly uniform heat flux boundary condition at the wall. This feature is troub-
lesome experimentally for conditions of low flow where the overall
heat loss may be a considerable proportion of the total heat dissipation of the element. In an attempt to allow for the axial distribution of heat loss, it was assumed that this loss was proportional to
the temperature difference between the wall and ambient air at a specified axial location with the implied constant of proportionality based
on the heat loss-rotational speed calibrations. The local bulk temperature was calculated on the basis of this axially distributed heat transfer assumption and, from this and the other readings, distributions
of the dimensionless parameters were obtained.

Further details of the apparatus and experimental programme are
given by Woods (1975). A comparison of the experimental data for air,
water and glycerol resulting from these investigations with theoretical predictions now follows. Table 3.2 summarises the range of experimental variables covered for these fluids.

For air the variation in the ratio of experimentally determined
Nusselt numbers under rotating and non-rotating conditions with the
product Re Ra$_\tau$ is shown in figure 3.26. Also shown in this figure is
the theoretical prediction from Woods and Morris (1974). Data for
both eccentricities are shown.

Bearing in mind that property effects other than the essential density variation in the analysis influence the real system, it may be
observed that the general trend of the experimental data follows the
prediction reasonably well. Generally, however, the theory tends to
overestimate the enhancement in heat transfer, although quantitative

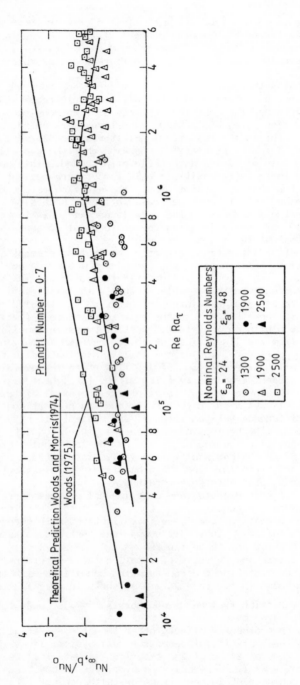

FIG. 3.26 COMPARISON OF THEORETICAL PREDICTIONS OF WOODS AND MORRIS (1974), WOODS (1975) WITH EXPERIMENTAL DATA FOR AIR.

Fluid	Air	Water	Glycerol
Prandtl Number	0.7	4 - 6	$10^3 - 10^4$
Reynolds Number	1300 - 2500	100 - 1200	0.1 - 2
L/d Ratio	34.7 - 69.3	48	48
Eccentricity (H/d)	24 - 28	24	24
Rotational Speed (rev/min)	0 - 1000	o - 300	0 - 300
Source	Woods (1975)	Morris (1964)	Morris (1964)

TABLE 3.2   RANGE OF VARIABLES COVERED BY MORRIS (1964)
AND WOODS (1975).

agreement is apparently better at higher Reynolds numbers. Significant improvements in the observed heat transfer are evident due to the rotation of the tube, for example, an improvement of 100 - 150 per cent at the higher rotational speeds.

A smooth line, about which the experimental data are evenly scattered, has been included in figure 3.26 to focus attention on an interesting feature. In the range $10^4 \leqslant Re\ Ra_\tau \leqslant 10^6$ the slope of the correlation is in good agreement with that predicted. The predicted improvement in heat transfer was about 25 per cent greater than the 'best fit' line drawn in the range $10^4 \leqslant Re\ Ra_\tau \leqslant 10^6$ and the data was scattered within 25 per cent of this line.

For $Re\ Ra_\tau$ values in excess of 10^6, the analysis still predicts sustained improvements in the heat transfer. However, figure 3.26 demonstrates that the actual enhancement in heat transfer does not appear to confirm this result. Indeed, the rate of enhancement is arrested with a subsequent tendency to decline. This feature will be discussed again later in this section.

Figure 3.27 shows the corresponding variation of heat transfer with rotation for water obtained from the data of Morris (1964). An experimentally determined reference Nusselt number correlation was not available from the original work of Morris (1964) owing to the nature of the original experiment. To overcome this, the heat transfer was expressed relative to stationary tube conditions evaluated with the laminar Seider-Tate (1936) correlation. These experiments with water covered a Prandtl number range between 4 - 6 and the theoretical predictions for both these Prandtl number values are shown in the figure.

Once again the results tend to confirm the predicted trends over a large portion of the experimental range covered. With water, the analysis tended to underestimate the enhancement by about 35 per cent in the range $10^3 \leqslant Re\ Ra_\tau \leqslant 5 \times 10^5$. Again, in the vicinity of $Re\ Ra_\tau = 10^6$, the enhancement in heat transfer levels off and subsequently declines in a manner similar to that found with air.

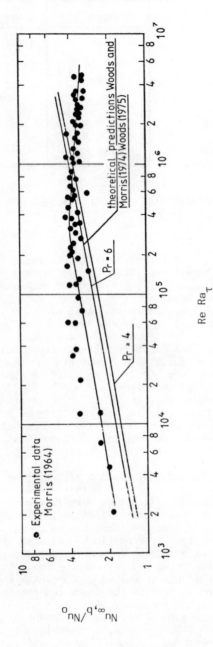

FIG. 3.27 COMPARISON OF THEORETICAL PREDICTIONS OF WOODS AND MORRIS (1974), WOODS (1975) WITH EXPERIMENTAL DATA FOR WATER FROM MORRIS (1964).

Figure 3.28 shows the data obtained with glycerol by Morris (1964). The range of Prandtl number covered with glycerol was between 10^3 – 10^4 and theoretical predictions from Woods and Morris (1974) are also shown for these values. All the data tend to fall within the bounds of these two predictions.

A number of possible explanations may be suggested for the discrepancy between the theoretical analysis and the experimental results. The dependency of viscosity on temperature is not taken into account in the present analysis. For air, the viscosity increases with temperature so that the secondary flow in the near-wall region will be suppressed in relation to the theoretical prediction. This could partially account for the observed overestimation of heat transfer. The converse argument applies to the water data.

Woods (1975) has postulated a possible explanation for the decline in heat transfer enhancement at relatively high values of the Re Ra_τ product noted with air and water. The explanation relates to the interaction of the axial variation of density in the flow direction with the centripetal acceleration. In this developed flow region the temperature of the fluid varies linearly in the direction of flow so that, at a given location in the cross stream direction, the upstream fluid is denser than the corresponding downstream location. If the predominant centripetal acceleration term is $\Omega^2 H$, this axial variation of density will give rise to an interaction likely to cause a circulation pattern similar to that shown in figure 3.29 to be superimposed on the axial velocity profile. Woods (1975) incorporated this modification into the numerical solution procedure and figure 3.30 shows the typical result which ensued. For a specified value of Re_p Ra_τ there was a tendency for the heat transfer to fall as the quotient Ra_τ/Re_p increased. In other words as the Rayleigh number increases at fixed pseudo Reynold number the enhancement in heat transfer diminishes. Although this mechanism does indicate the trends noted with the experiments undertaken with air and water precise quantitative agreement did not occur. This interesting effect is worthy of greater detailed investigation in the future.

The present problem is analogous to that of a horizontal tube which is influenced by the earth's gravitational field. This is particularly true at large values of eccentricity. For the horizontal tube situation with heated developed laminar flow Siegworth et al (1969) have shown that the axial velocity profile is unaffected by the cross stream gravitational buoyancy at high values of Prandtl number. This led Woods (1975) and Woods and Morris (1981) to examine the high Prandtl number asymptotic solution for the rotating tube situation.

For high Prandtl number values these authors confirmed that rotation did not influence the axial velocity profile with the consequential effect that the friction factor would equally become independent of rotation. This was, of course, already apparent from the trends shown in figure 3.17. Further it was demonstrated that the influence of Prandtl number on the solution could be uniquely taken into account by the parameter Re_p Ra_τ Pr instead of the previously used product Re_p Ra_τ. Figure 3.31 shows the asymptotic solution for heat transfer

92

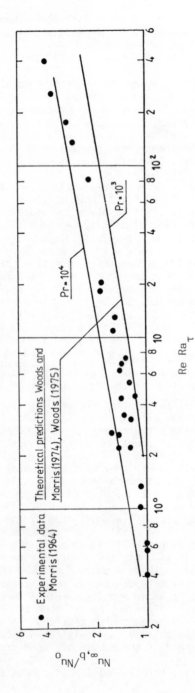

FIG. 3.28 COMPARISON OF THEORETICAL PREDICTIONS OF WOODS AND MORRIS (1974), WOODS (1975) WITH EXPERIMENTAL DATA FOR GLYCEROL FROM MORRIS (1974).

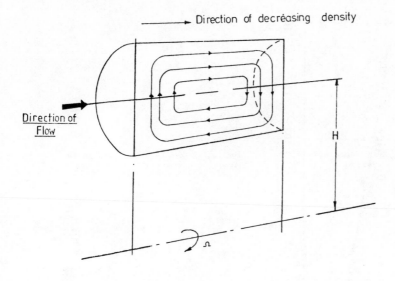

FIG. 3.29   EXPECTED CIRCULATION DUE TO AXIAL DENSITY
VARIATION. WOODS (1975).

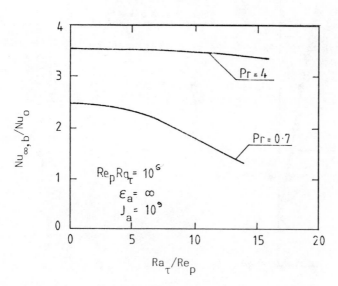

FIG. 3.30   EFFECT OF AXIAL DENSITY CORRECTION ON HEAT
TRANSFER. WOODS (1975).

94

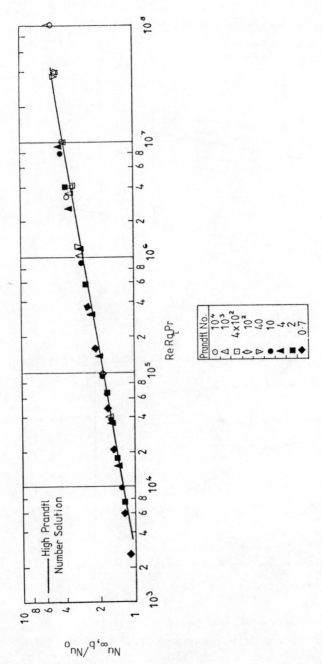

FIG. 3.31   COMPARISON OF THEORETICAL RESULTS WITH ASYMPTOTIC HIGH PRANDTL NUMBER
SOLUTION. WOODS (1975), WOODS AND MORRIS (1981).

which resulted and also the individual predictions obtained at various Prandtl number values. It is interesting to note that, although developed on the basis of the high Prandtl number asymptote, this solution gives close agreement with Prandtl number values down to about unity. This led Woods and Morris (1981) to propose the theoretically generated heat transfer correlation.

$$\frac{Nu_{\infty,b}}{Nu_o} = 0.262[Re\ Ra_\tau\ Pr]^{0.173} \tag{3.124}$$

valid in the range

$$4 \times 10^3 \leqslant Re\ Ra_\tau\ Pr \leqslant 10^8$$

$$0.7 \leqslant Pr \leqslant 10^4$$

The experimentally available data for air, water and glycerol is compared with equation (3.124) in figure 3.32. Although data scatter is inevitable with this type of experiment the data follows the trend suggested by the theoretical model reasonably well. This confirms that the simplified method of data correlation suggested by the high Prandtl number solution is a useful technique for data presentation in this case.

This section will now be concluded with a brief discussion of the capability of the numerical solution for developed laminar flow to predict details of the velocity and temperature fields. Owing to the nature of the flow geometry it is difficult to make detailed measurements of the flow and temperature fields. Indeed this has not yet been attempted as far as the present author is aware. However Mori et al (1966) have made measurements of axial velocity distributions in a heated horizontal tube subjected to the earth's gravity. Temperature profiles have also been reported by these authors for this analogous problem. Figures 3.33 and 3.34 show comparisons made by Woods (1975) between the numerical procedure and the analogous experiments of Mori et al (1966). Agreement is encouragingly good.

The asymmetric temperature profile produced as a consequence of rotation results in a corresponding circumferential variation in local heat flux at the tube wall. Sakamoto and Fukui (1971) assessed this circumferential variation of flux using the well known analogy between heat and mass transfer. These authors measured the rate of naphthalene transfer from a coated inner tube wall for a variety of rotational conditions. Figure 3.35 shows, for a typical set of operating conditions, the variation of mass transfer detected. For convenience the ordinate in figure 3.35 has been normalised in relation to the maximum value which occurs at the outermost radial location ($\theta = 0$ or 2π). Also shown in figure 3.35 for the same analogous condition is the variation of normalised heat flux resulting from the numerical solution. Generally the agreement is good with the correct pattern of behaviour being predicted.

96

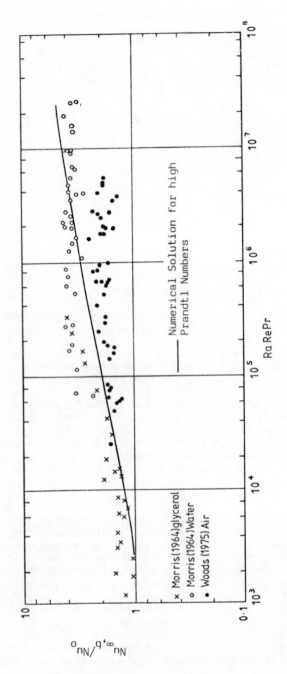

FIG. 3.32 COMPARISON BETWEEN THEORETICAL HIGH PRANDTL NUMBER SOLUTION AND
EXPERIMENTAL DATA.

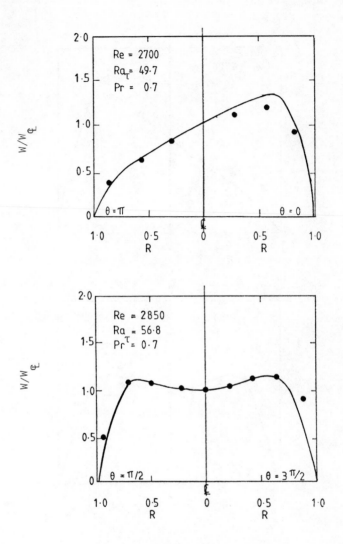

FIG. 3.33   COMPARISON OF AXIAL VELOCITY PREDICTION
WITH EXPERIMENTAL DATA FOR ANALOGOUS
HORIZONTAL TUBE.   MORI ET AL (1966),
WOODS (1975).

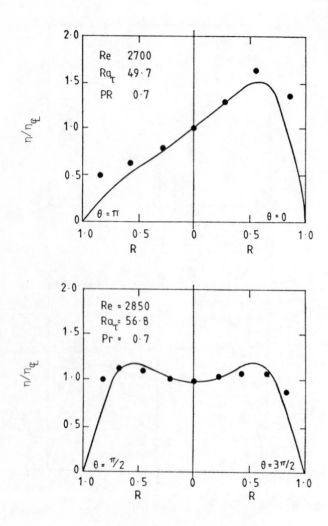

FIG. 3.34 COMPARISON OF TEMPERATURE PREDICTION
WITH EXPERIMENTAL DATA FOR ANALOGOUS
HORIZONTAL TUBE. MORI ET AL (1966),
WOODS (1975).

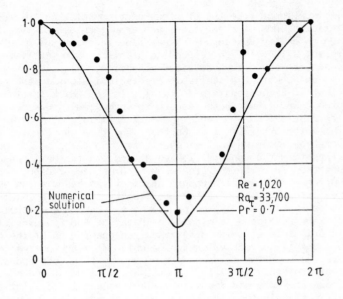

FIG. 3.35　A COMPARISON BETWEEN THE NUMERICAL
SOLUTION FOR THE CIRCUMFERENTIAL HEAT
TRANSFER DISTRIBUTION AND THE RESULTS
OF SAKAMOTO AND FUKUI (1971) FOR THE
SUBLIMATION OF NAPHTHALENE.

3.4 Laminar Heat Transfer in the Entrance Region

It is evident from the previous sections of this chapter that a substantial amount of theoretical and experimental work has been undertaken on heated developed laminar flow in tubes which rotate about a parallel axis. The corresponding problem in the entry region has not, to date, received such detailed attention.

The only currently available theoretical study in the entry region is that of Skiadaressis and Spalding (1976) which treats the uniformly heated tube. These authors used the same basic conservation equations as Morris (1965b) and Woods and Morris (1974) but modified to include axial velocity gradients. Further, the flow was treated as boundary layer-type implying that viscous action in the axial direction was omitted from the conservation equations. A further boundary layer assumption involved the treatment of pressure gradient terms in the equations. The axial pressure gradient was based on an average cross stream pressure which was assumed constant at any axial location. This assumption permitted the use of a marching numerical solution technique developed originally by Patankar and Spalding (1972).

In principle a finite difference representation of the conservation equations is solved by forward marching from an upstream cross section, where the solution is deemed to be known, to a corresponding downstream station. Initially an estimate is made of the new downstream average pressure and the axial velocity calculated. This velocity distribution is then checked against the specified mass flow rate along the duct and corrections made as required to the assumed average pressure.

This is followed by an evaluation of the cross stream velocities using an initially assumed cross stream pressure profile with subsequent corrections made in order to ensure that continuity is satisfied at all control volumes surrounding each nodal point selected in the flow field. When the velocity field has been satisfactorily determined the finite difference form of the energy equation is solved to give the temperature distribution.

Skiadaressis and Spalding (1976) elected to use the Rossby number, Ro, to characterise the Coriolis effect. Unfortunately this paper has concentrated mainly on demonstrating that the solution procedure is capable of solving the problem. A detailed study of the way in which the controlling parameters influence the heat transfer mechanism is not presented but, nevertheless, some interesting results are forthcoming. Figure 3.36 illustrates the typical manner in which rotation affects the distribution of local heat transfer plotted as a local Nusselt number, Nu_z, which is defined in terms of the local wall to bulk fluid temperature difference. These local distributions are based on the assumption that the fluid enters the heated tube with a flat axial velocity profile and with uniform temperature. A number of the local Nusselt number curves shown in figure 3.36 have a tendency to 'ripple' in the downstream direction. It is difficult to judge whether this is a real physical effect or indeed some numerically induced oscillation in the solution procedure. As we shall see later there has been no evidence of this axial oscillation in experimentally measured distributions of local Nusselt number. The fully

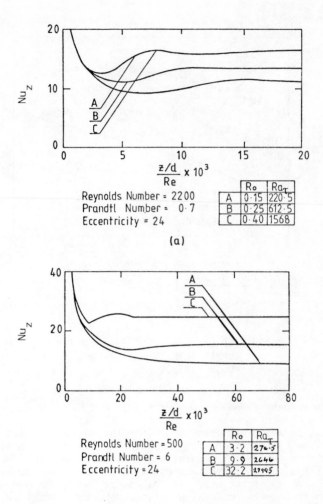

Reynolds Number = 2200
Prandtl Number = 0·7
Eccentricity = 24

	Ro	RaT
A	0·15	220·5
B	0·25	612·5
C	0·40	1568

(a)

Reynolds Number = 500
Prandtl Number = 6
Eccentricity = 24

	Ro	RaT
A	3·2	274·5
B	9·9	2646
C	32·2	17995

FIG. 3.36   PREDICTIONS OF ENTRY REGION HEAT TRANSFER
SKIADARESSIS AND SPALDING (1976).

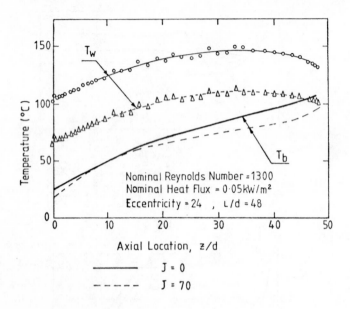

Axial Location, z̄/d

———————— J = 0

— — — — J = 70

FIG. 3.37   TYPICAL EFFECT OF ROTATION ON WALL AND BULK
TEMPERATURE IN THE ENTRANCE REGION.
WOODS (1975).

developed values of Nusselt number were found to be in good agreement
with those presented by Woods and Morris (1974).

The only detailed measurements of developing laminar heat transfer
in this flow system are due to the investigation of Woods (1975) using
air. Even so the data is sparse and the point is made by Woods (1975)
that this type of experiment is particularly difficult to set up and
control. The apparatus used was that described by Woods and Morris
(1974) in their work on developed flow, details of which feature in
the last section of this book. Because the external heat losses from
the rotating test section tend to be a large proportion of the overall
energy dissipated in the electrical heaters used, it is difficult to
maintain a truly uniform heat flux thermal boundary condition experimentally. Woods (1975) used an elaborate energy accounting procedure
to correct for this effect and figure 3.37 illustrates the typical way
in which the wall temperature of the tube and the corresponding fluid
bulk temperature respond to rotation.

The curves shown in figure 3.37 correspond to a nominal Reynolds
number of 1300 and a nominal heat flux of 0.05 kW/m². The dramatic
reduction in wall temperature brought about by rotation is an immediate indication of the improved heat transfer. Note that the departure
of the bulk temperature from a linear axial variation reflects the

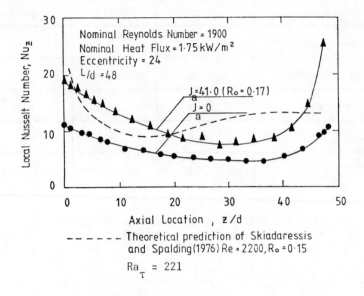

FIG. 3.38   TYPICAL AXIAL VARIATION OF LOCAL NUSSELT
NUMBER. WOODS (1975).

problem of creating a truly uniform heat flux condition at the wall
at fairly low rates of coolant flow.

Figure 3.38 typifies the development of laminar heat transfer noted
by Woods (1975) and the general enhancement of local Nusselt number
due to rotation is very significant. The sharp increase in Nusselt
number at the exit region of the tube is due to end loss effects which
are inevitably present with forced convection experiments in ducts.
Although not strictly comparable figure 3.38 also includes the prediction for Re = 2200, Ro = 0.15, Ra = 221 from the work of Skiadaressis and Spalding (1976). It is apparent that the general trend of
agreement is not good in the developing region and there is a clear
need for more detailed experimental and theoretical study. A particularly useful experimental investigation would involve a search for the
'rippling' in the axial Nusselt number distribution highlighted by the
theoretical studies of Skiadaressis and Spalding (1976).

Average heat transfer data in relatively short tubes have been reported by Sakamoto and Fukui (1971) and Morris and Woods (1978).

Sakamoto and Fukui (1971) used air and transformer oil as the test
fluids and also undertook some mass transfer experiments whereby naphthalene sublimation from the walls of the rotating tube to air was
measured. Table 3.3 gives the range of variables covered in their experiments.

Figure 3.39 shows the influence of rotation on the heat transfer to
air reported by Sakamoto and Fukui (1971). Although precise details
of the experiment are not presented in their paper it is apparent that

Fluid	Air	Oil	Naphthalene
Prandtl Number	0.72	400	2.62
Reynolds Number	$490 - 1.74 \times 10^4$	$162 - 2 \times 10^3$	$535 - 3.19 \times 10^4$
L/d Ratio	20.4	20.0	24.0
Eccentricity (H/d)	10.0	10.0	10.0
Rotation Speed (rev/min)	420 - 2700	522 - 2700	528 - 2700

TABLE 3.3   SYNOPSIS OF EXPERIMENTAL VARIABLES COVERED
BY SAKAMOTO AND FUKUI (1971)

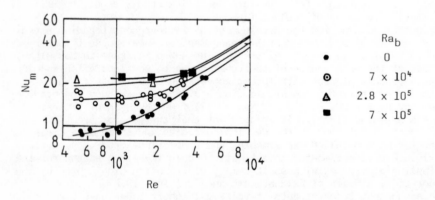

FIG. 3.39   EFFECT OF ROTATION ON MEAN HEAT TRANSFER
FOR AIR.   SAKAMOTO AND FUKUI (1971).
(H/d = 10, L/d = 20.4).

experiments were arranged to control the rotational Rayleigh number over a range of Reynolds number values. It is evident from this fig-ure that the rotation is producing a marked improvement in heat trans-fer in the laminar flow range.

Sakamoto and Fukui (1971) proposed that all their experimental data could be correlated in terms of the rotational Rayleigh number, Ra_b, and the Graetz number, Gz. The equation proposed had the structure

$$\frac{Nu_m}{Nu_o} = \left[1 + 0.03 \frac{Ra_b^{0.75}}{Gz} \right]^{1\cdot3} \tag{3.125}$$

where

$$\left. \begin{array}{l} Gz = \frac{\pi}{4} \, Re \, Pr \, \frac{d}{L} \\[3mm] \text{and } Ra_b = \dfrac{H\Omega^2\beta\rho^2 c_p d^3(T_w - T_b)}{\mu K} \end{array} \right\} \tag{3.126}$$

Note that the definition of the rotational Rayleigh number, Ra_b, used by these authors involves the tube diameter and the wall to bulk fluid temperature difference.

Equation (3.125) was established with data covering the range $162 \leqslant Re \leqslant 2700$, $Ra_b \leqslant 2 \times 10^7$.

One interesting feature of the work of Sakamoto and Fukui (1971) in-volved the assessment of circumferential variations in heat flux around the periphery of the rotating tube. This was done via the mass transfer/heat transfer analogy. The authors do not state clearly how Rayleigh number is evaluated in the mass transfer studies but the cir-cumferential trends were similar to those resulting from the developed analysis of Morris and Woods (1974) and reported in Woods and Morris (1981). Figure 3.35 has already illustrated this effect.

The experimental data for air, water and glycerol available from the studies of Woods (1975) and Morris (1964, 1968) have been re-evaluated on a mean basis for the tubes concerned and these results are compared with the proposal of Sakamoto and Fukui (1971) in figure 3.40. Equa-tion (3.125) does not correlate this additional data particularly well and this led Morris and Woods (1978) to seek an alternative empirical correlation for their own data. In their paper Morris and Woods (1978) deal firstly with turbulent flow where the effect of buoyancy, they argue, is likely to be less important than the Coriolis effect. Their results from the analysis of turbulent experiments led them to consider a similar form of correlation for laminar flow and this condition will now be discussed. Details of turbulent flow studies for this rotating geometry will be dealt with in the next chapter.

Morris and Woods (1978) reported data for two test sections. For H/d = 48 and L/d = 34.65, figure 3.41 shows the typical variation of mean Nusselt number, Nu_m, for two nominal Reynolds number values.

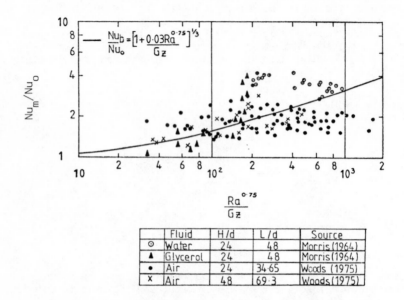

	Fluid	H/d	L/d	Source
⊙	Water	24	48	Morris (1964)
▲	Glycerol	24	48	Morris (1964)
●	Air	24	34·65	Woods (1975)
×	Air	48	69·3	Woods (1975)

FIG. 3.40 COMPARISON OF DATA FROM MORRIS (1964) AND WOODS (1975) WITH PROPOSED CORRELATION FOR MEAN LAMINAR HEAT TRANSFER FROM SAKAMOTO AND FUKUI (1971).

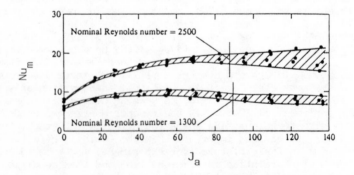

FIG. 3.41 TYPICAL EFFECT OF ROTATION ON MEAN LAMINAR HEAT TRANSFER FROM MORRIS AND WOODS (1978). (H/d = 48, L/d = 34.65).

These authors attempted to correlate the data in terms of the rotational Reynolds number, J_a, together with the Reynolds number. Since the data was obtained with air, for which there was not a significant change in the Prandtl number, no attempt was made to explicitly include the Prandtl number in their correlations.

The bandwidth shown in figure 3.41 bounds tests undertaken with different heating rates and in this respect therefore includes a buoyancy effect to some extent. The heat transfer enhancement tends to flatten off at the higher rotational speeds, this effect being consistent with the observations made on developed heat transfer earlier in this chapter.

Figure 3.42 compares the effect of rotational Reynolds number on mean heat transfer for both test sections studied. The data is shown for two nominal Reynolds number values of 1900 and 2500 respectively. For laminar flow it was found that a single correlating equation for both test sections could not be postulated. (This had been possible for the case of turbulent flow as will be shown in the next chapter). Morris and Woods (1978) made the following proposals for correlating their mean heat transfer data

Case A: H/d = 48 , L/d = 34.65

$$Nu_m = 0.016 \ Re^{0.78} \ J_a^{0.25} \tag{3.127}$$

Case B: H/d = 96 , L/d = 69.30

$$Nu_m = 0.013 \ Re^{0.78} \ J_a^{0.25} \tag{3.128}$$

These equations, which were based on experiments with values of rotational Reynolds number upto about 150, are compared with the experimental data in figure 3.43.

The correlation is not particularly good at the higher rotational speeds where the influence of buoyancy will become more important. There is insufficient data currently available to take fully into account the complex interactions between Coriolis acceleration and centripetal buoyancy.

3.5   Recommendations for Design Purposes

It was stated in Chapter 1 that the strategic aim of this monograph was to review published work in the general field of fluid mechanics and heat transfer in rotating cooling channels and to make recommendations which may be used in the design role. This section accordingly itemises the main conclusions and recommendations for the case of laminar flow in tubes which rotate about a parallel axis.

Flow Resistance with Unheated Flow

When the flow is laminar and unheated rotation does not affect resistance to flow once the flow field has been established. For chan-

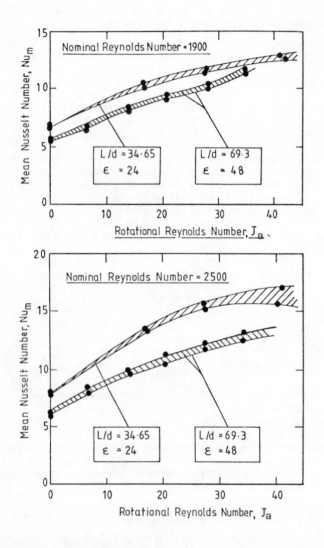

FIG. 3.42   EFFECT OF GEOMETRY ON MEAN LAMINAR HEAT
            TRANSFER.   MORRIS AND WOODS (1978).

109

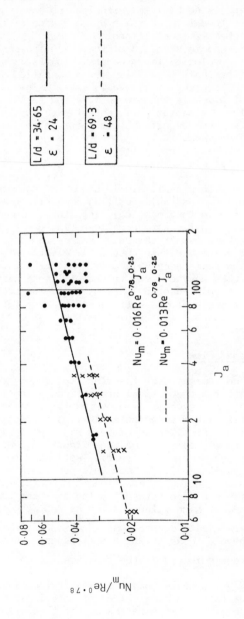

FIG. 3.43 COMPARISON OF LAMINAR FLOW DATA WITH PROPOSED CORRELATIONS FOR MEAN
HEAT TRANSFER. MORRIS AND WOODS (1978).

nels having a large length/diameter ratio, where fully developed flow will dominate the regime of interest, the resistance offered to flow may therefore be estimated from the customary relationship between friction factor and Reynolds number. Nevertheless no information currently exists with which to determine the precise length of tube necessary to ensure established flow conditions.

For short tubes, where the interaction of the Coriolis acceleration and an axially developing velocity field occurs, flow resistance will be increased over the stationary tube value for a specified Reynolds number. The only information currently available with which to estimate this effect is that given in equations (3.7) and (3.14). These equations may be used for estimation purposes but care should be exercised not to extrapolate these results beyond the range of variables over which they were determined.

Flow Resistance with Heated Flow

For heated flow the only data available is based on theoretical studies with the assumption of established flow. No direct experimental data has been reported in the technical literature. The theoretical predictions of heat transfer with developed flow due to Woods and Morris (1974) appear to be the most versatile and reliable over a wide range of Prandtl number values. It is equally likely therefore that their predictions of flow resistance will give the best theoretically based predictions for tubes having large values of length/diameter ratio. Thus the information presented in figure 3.17 is recommended for the estimation of pressure drop. Until specific data for short tubes becomes available figure 3.17 may also be used for a ranging calculation of resistance.

Heat Transfer in Developed Flow

The review of developed laminar heat transfer studies presented earlier demonstrates that the theoretical predictions given in figure 3.16 are well supported, in the main, by experimental studies. Although subject to quantitative discrepancies they do correctly demonstrate the effect of all controlling parameters expected to influence the problem. Further, the outcome of the high Prandtl number asymptote suggests that heat transfer may be readily estimated for design purposes by the use of equation (3.124) and this procedure is recommended for tubes having a relatively large length/diameter ratio.

For air and water there has been a tendency for the heat transfer enhancement to be reduced at high values of the Re Ra_τ product. Care should be exercised therefore to ensure that equation (3.124) is not used at Re Ra_τ values beyond those at which heat transfer enhancement appears to flatten off as indicated by the experimental data shown in figures 3.26 and 3.27.

The work of Woods (1975) suggests that the effect of Coriolis acceleration will become significant at low values of rotational Reynolds numbers and of eccentricity. After a review of his theoretical results Woods (1975) suggests that the Coriolis effect on developed heat

transfer will be small provided $J\varepsilon_a < 0.025$. In a typical electrical
machine application $J\varepsilon_a$ is usually less than 0.0001 so that Coriolis
effects should not be significant.

Heat Transfer in the Entrance Region

The parameters which govern heat transfer in the entry region have
not been comprehensively reported theoretically or substantiated experimentally. Mean heat transfer for gas-like flows may be estimated
from the empirical equations (3.127) and (3.128) provided their range
of validity is not severely violated. Eccentricity does appear to
have a small but noticeable effect. These equations are useful for
electrical machine cooling applications.

For a wider range of Prandtl number values, equation (3.125) may also be used to estimate heat transfer. The use of developed values of
heat transfer enhancement may always be used as a guide to cooling
performance.

CHAPTER 4

TURBULENT FLOW AND HEAT TRANSFER IN

CIRCULAR-SECTIONED TUBES WHICH

ROTATE ABOUT A PARALLEL AXIS.

4.1 Introduction

    This chapter deals with the same flow geometry as the previous chap-
ter but considers specifically the case of a notionally turbulent flow.
The only data reported for isothermal flow resistance is that of
Morris (1981) discussed in the previous chapter so that, of necessity,
this chapter is devoted entirely to aspects of theoretical and experimental investigations of heated flows.

4.2 Heated Flows in Circular-Sectioned Ducts

4.2.1 Opening Remarks

    A number of theoretical and experimental studies have been reported
for conditions of turbulent flow and these will be discussed in this
section using a similar presentational format to that used for laminar
flow in the previous chapter.

4.2.2 Theoretical Studies of Developed Turbulent Flow

    The first theoretical attempt to study turbulent heated flow in
this sytem was reported by Nakayama (1968). In essence Nakayama
(1968) considered developed flow in a uniformly heated tube so that,
with the exception that the flow was turbulent, the problem was posed
in an identical fashion to that discussed in section 3.3.2 of the
last chapter. In fact this author used an identical solution technique to that used for the laminar counterpart reported by Mori and
Nakayama (1967). Thus an effective momentum integral method of solution was used with an assumed core flow region comprising a strong
secondary flow together with a relatively thin boundary layer in the
near-wall vicinity. The solution is claimed to be valid for fluids
having a Prandtl number of about unity or greater. Because the general procedure has been described in detail in the last chapter only
the main features and eventual results will be discussed here.
    The main additional physical effect which had to be included in the
analysis was the incorporation of the turbulent fluctuations of velocity and temperature into the basic conservation equations of momentum

and energy.

For gas-like fluids, with Prandtl number values close to unit, Nakayama (1968) developed an expression for the bulk Nusselt, $Nu_{\infty,b}$ which, in terms of the nomenclature of the present text, may be written as

$$\frac{Nu_{\infty,b}}{Nu_o} = \frac{1.367\ Pr^{2/3}}{(Pr^{2/3} - 0.050)}\left[1 + \frac{0.0286}{X^{1/5}}\right] X^{1/20} \tag{4.1}$$

where

$$X = \frac{Ra_{\tau}^{1.818}}{Re^{2.273}\ Pr^{0.606}} \tag{4.2}$$

and the bulk Nusselt number, Nu_o, for the stationary tube is calculated from

$$Nu_o = 0.038\ Re^{3/4}\ Pr^{1/3} \tag{4.3}$$

For liquids Nakayama's analysis resulted in the following expression for heat transfer enhancement

$$\frac{Nu_{\infty,b}}{Nu_o} = 1.428\left[1 + \frac{0.0144}{X^{1/6}}\right] X^{1/30} \tag{4.4}$$

where in this case

$$X = \frac{Ra_{\tau}^{2.308}}{Re^{3.231}\ Pr^{0.923}} \tag{4.5}$$

and

$$Nu_o = 0.023\ Re^{0.8}\ Pr^{0.4} \tag{4.6}$$

Figure 4.1 illustrates the level of enhancement in heat transfer which is implied by the analysis of Nakayama (1968) for gas-like and liquid-like fluids. As might have been anticipated, with turbulent flow the influence of rotational buoyancy is not as marked as that found with laminar flow. The influence of Coriolis acceleration was found to be negligibly small for developed turbulent flow.

Separate equations for the impediment to flow resistance were developed for gas-like and liquid-like flows. Thus for gases Nakayama (1968) proposed

$$\frac{C_{fR}}{C_{fo}} = 1.292\left[1 + \frac{0.0329}{X^{1/5}}\right] X^{1/20} \tag{4.7}$$

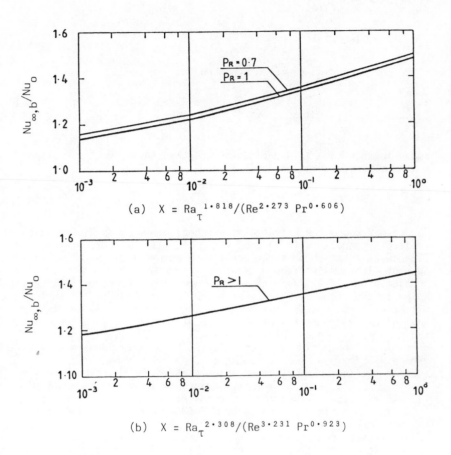

(a) $X = Ra_\tau^{1\cdot818}/(Re^{2\cdot273}\ Pr^{0\cdot606})$

(b) $X = Ra_\tau^{2\cdot308}/(Re^{3\cdot231}\ Pr^{0\cdot923})$

FIG. 4.1   THEORETICAL PREDICTIONS FOR DEVELOPED
TURBULENT HEAT TRANSFER. NAKAYAMA (1968).

where X is defined by equation (4.2) and the stationary tube Blasius friction factor C_{fo}, by

$$C_{fo} = 0.316 \ Re^{-1/4} \qquad (4.8)$$

The corresponding equation proposed for liquids is

$$\frac{C_{fR}}{C_{fo}} = 1.434 \left[1 + \frac{0.0157}{X^{1/6}} \right] X^{1/30} \qquad (4.9)$$

where in this case X is defined by equation (4.5) and the stationary tube reference condition by

$$C_{fo} = 0.184 \ Re^{-1/5} \qquad (4.10)$$

Figure 4.2 shows the effect of rotational buoyancy on flow resistance implied by equations (4.7) and (4.9). As with heat transfer the influence of rotation on turbulent flow pressure loss is not as severe as that with laminar flow.

Majumdar et al (1977) extended the numerical method used for developing laminar flow by Skiadaressis and Spalding (1976), to include the effect of turbulence. This necessitates a mathematical model by which the influence of the turbulent nature of the flow may be included in the conservation equations and this will now be briefly outlined. For convenience the motion will be referred to a Cartesian frame of reference.

The so-called boundary layer assumption in the momentum equations involves the omission of velocity gradients in the main direction of flow in relation to their cross stream counterparts. For the turbulent flow condition considered by Majumdar et al (1977) they linked the boundary layer assumption to the assumption that the normal and shear stresses could be expressed in terms of a laminar flow mathematical structure but with an effective viscosity term, μ_e, to account for the additional turbulent stresses. With these assumptions the normal and shear stresses terms presented in Tables 2.3 and 2.4 of Chapter 2 can be combined to give, for turbulent flow, the results shown in Table 4.1. The velocity terms shown in Table 4.1 are time-average values and the flow is taken to be mainly along the z-axis.

The effective viscosity, μ_e, was calculated from a version of the two-equation model of turbulence proposed by Harlow and Nakayama (1967) modified by Launder and Spalding (1974). In this turbulence model the effective viscosity is written as

$$\mu_e = \mu + C \ \rho \ k^2/\underline{\varepsilon} \qquad (4.11)$$

where k is the turbulent kinematic energy, $\underline{\varepsilon}$ is the rate of turbulent dissipation and C is a constant (taken to be 0.09). The turbulence

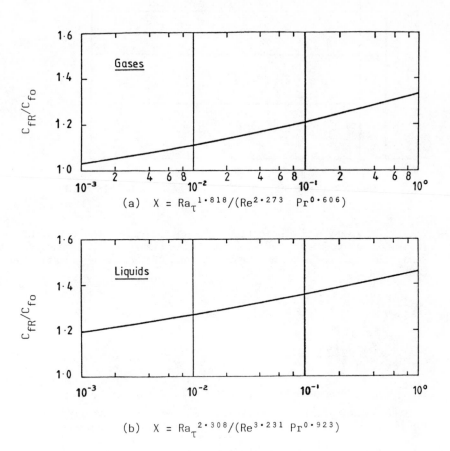

(a) $X = Ra_\tau^{1 \cdot 818}/(Re^{2 \cdot 273} Pr^{0 \cdot 606})$

(b) $X = Ra_\tau^{2 \cdot 308}/(Re^{3 \cdot 231} Pr^{0 \cdot 923})$

FIG. 4.2   THEORETICAL PREDICTIONS FOR DEVELOPED
TURBULENT FLOW RESISTANCE. NAKAYAMA (1968).

		Normal and Shear Stress Terms for Turbulent Boundary Layer Flow.
Coordinate Direction	X	$-\dfrac{\partial P}{\partial x} + 2\dfrac{\partial}{\partial x}\left[\mu_e\dfrac{\partial u}{\partial x}\right] + \dfrac{\partial}{\partial y}\left[\mu_e\left(\dfrac{\partial v}{\partial x} + \dfrac{\partial u}{\partial y}\right)\right]$
	Y	$-\dfrac{\partial P}{\partial y} + 2\dfrac{\partial}{\partial y}\left[\mu_e\dfrac{\partial v}{\partial y}\right] + \dfrac{\partial}{\partial x}\left[\mu_e\left(\dfrac{\partial v}{\partial x} + \dfrac{\partial u}{\partial y}\right)\right]$
	Z	$-\dfrac{\partial P}{\partial z} + \dfrac{\partial}{\partial x}\left[\mu_e\dfrac{\partial w}{\partial x}\right] + \dfrac{\partial}{\partial y}\left[\mu_e\dfrac{\partial w}{\partial y}\right]$

TABLE 4.1 TURBULENT BOUNDARY LAYER FLOW STRESSES

kinematic energy and dissipation rate are linked by two differential equations which are solved simultaneously with the other conservation laws. In this respect the turbulent form of the energy equation was taken to be

$$\frac{\partial}{\partial x}(\rho u T) + \frac{\partial}{\partial y}(\rho v T) + \frac{\partial}{\partial z}(\rho w T)$$

$$= \frac{\partial}{\partial x}\left[\frac{\mu_e}{Pr_t}\frac{\partial T}{\partial x}\right] + \frac{\partial}{\partial y}\left[\frac{\mu_e}{Pr_t}\frac{\partial T}{\partial y}\right] + \frac{\partial}{\partial z}\left[\frac{\mu_e}{Pr_t}\frac{\partial T}{\partial z}\right] \tag{4.12}$$

where Pr_t is the turbulent Prandtl number (taken to be 0.9).

Although only limited results for developed flow are presented in the paper of Majumdar et al (1977) the technique appears to be capable of predicting the trends for hydrodynamic and thermal development. A comparison of their prediction for developed turbulent heat transfer with that of Nakayama (1968) is shown in figure 4.3 for air. The prediction for air is somewhat higher than that of Nakayama (1968). The influence of Coriolis acceleration on developed flow was again shown to be small.

4.2.3 Experimental Studies of Developed Turbulent Heat Transfer and Comparison with Theoretical Predictions.

Only two studies have been reported in sufficient detail to permit a comparison of theoretical predictions with experimental evidence for developed turbulent flow. These are due to Woods (1975) and Nakayama and Fuzioka (1978) and give data for air and water respectively.

The data of Woods (1975) was collected using the same apparatus as that described in the previous chapter for laminar flow and Table 4.2

Reference	Fluid	L/d	H/d	Re	Ra_τ	J_a	Pr
Woods (1975)	Air	34.7	24	$5 \times 10^3 - 2 \times 10^4$	$2 \times 10^2 - 4 \times 10^3$	15 – 150	0.7
Nakayama and Fuzoika (1978)	Water	41.7	27.3	$5 \times 10^3 - 2 \times 10^4$	$10^5 - 10^6*$	4500*	5.5 – 6.7
	Water	180	?	$7 \times 10^3 - 3.7 \times 10^4$	$10^5 - 10^7*$	4500*	4.0 – 4.5

* Estimated from data in source references.

TABLE 4.2   RANGE OF EXPERIMENTAL VARIABLES AVAILABLE FOR DEVELOPED TURBULENT HEAT TRANSFER

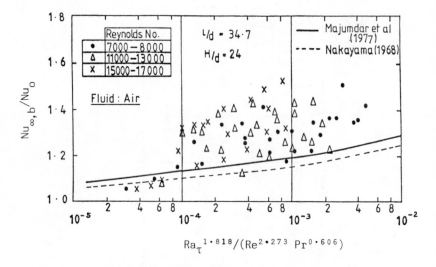

FIG. 4.3 COMPARISON OF THEORETICAL PREDICTIONS FOR
TURBULENT DEVELOPED FLOW WITH EXPERIMENTAL
DATA OF WOODS (1975).

illustrates the range of variables tested. Figure 4.3 compares the
experimental results obtained for air with the theoretical predictions
of Nakayama (1968) and Majumdar et al (1976). In general terms the
comparisons are not good with a tendency of the experimental data to
be significantly higher than that predicted. The theoretical prediction of Majumdar et al (1976) which is based on a good, theoretical
differential model, is only in reasonable accord with the lower bound
of the experimental data. This discrepancy is not necessarily a crit-
icism of the predictions or the experimental data but probably reflects the relative importance of the Coriolis acceleration and centripetal buoyancy. Also, as will be demonstrated later in this chapter,
the influence of upstream delivery ducting conditions the velocity
profile at the immediate entry plane of the test section with consequential affect on the development length. This affect is also coupled as demonstrated in Chapter 2 to the Coriolis acceleration.
Nakayama and Fuzoika (1978) made experimental observations in the
cooling channels of a prototype generator and also a simulated labora-
tory apparatus using water as the test fluid. This is the only recorded data for fluids other than air which explicitly quotes developed heat transfer results. Their paper does not present sufficient
details of the apparatus to give a full description here but Table
4.2 shows the salient range of variables studied. In some instances
estimates have been made due to limited information. Figure 4.4 compares their experimental findings with the prediction of Nakayama
(1968) given by equation (4.4). The agreement is very encouraging
and far superior to that for gas-like flows. It could be argued that
the centripetal buoyancy has a much stronger effect with the water

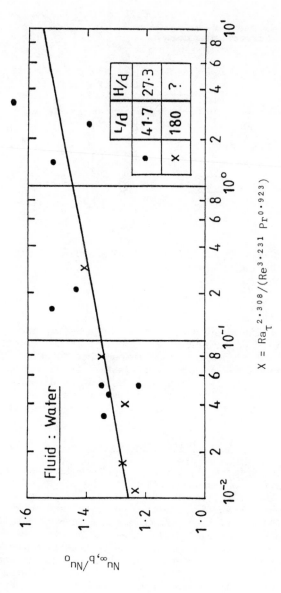

$$X = Ra_\tau^{2.308} / (Re^{3.231} \, Pr^{0.923})$$

FIG. 4.4   COMPARISON OF THEORETICAL PREDICTION FOR TURBULENT DEVELOPED FLOW FROM NAKAYAMA (1968) WITH EXPERIMENTAL DATA FOR WATER FROM NAKAYAMA AND FUZIOKA (1978).

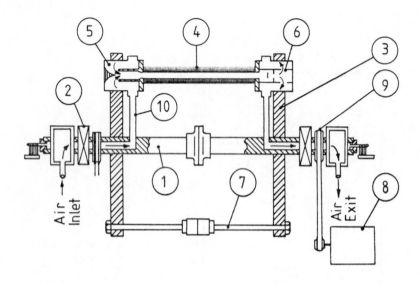

FIG. 4.5   SCHEMATIC ARRANGEMENT OF APPARATUS USED BY
HUMPHREYS (1966).

and this is confirmed by the rotational Rayleigh numbers shown in
Table 4.2.   Even so there is also a much larger rotational Reynolds
number operating with the water experiments.   As a set of working
figures for comparative purposes the rotational Rayleigh numbers for
water are about one thousand times greater than those studied for air
whereas the rotational Reynolds numbers for water are only about fifty
times those for air.   It is clear that more data is required with
which to unravel the complex interaction between Coriolis and centripetal effects in order to ensure a truly established flow experimental
regime.

4.2.4   Turbulent Heat Transfer in the Entrance Region.

Humphreys (1966) experimentally investigated turbulent heat transfer.   The apparatus is shown schematically in figure 4.5 and consisted
of a built up rotor shaft (1) mounted between self aligning bearings
(2).   Two steel support arms (3) permitted the test section (4) together with entry (5) and exit (6) chambers to be mounted parallel to
the shaft axis with an eccentricity of either 152.4 mm or 304.8 mm.
Weights (7) fitted diametrically opposite the test section facilitated
balance of the rotor which was driven by means of a variable speed
electric motor (8) via a pulley system (9).   Air was circulated
through the test section via internal passages in the rotor shaft and
radial connecting tubes (10) by means of a blower.   The air flow rate
was measured with a bell mouth flowmeter.

The test section was made from brass tube nominally 304.8 mm long with two alternative bore diameters of 25.4 mm and 6.4 mm respectively. The tube surfaces were electrically heated by means of spirally wound resistance wire with electrical insulation between the tube and wire achieved using glass fibre tape. The test section wall temperature was measured with copper/constantan thermocouples along its entire length and also the air temperature in the inlet and exit chambers. All thermocouple signals were taken from the rotor using a miniature slip ring assembly (11). The test section was lagged to minimise external heat loss.

Table 4.3 summarises the range of variables tested and the salient features of the study are reported by Humphreys et al (1967). Figure 4.6 shows the typical manner in which rotation influenced the local Nusselt number, Nu_z, along the test section. The local Nusselt number Nu_z, is defined in terms of the locally prevailing heat flux and wall to fluid bulk temperature difference. The increase in local heat transfer with increases in rotation speed is clearly evident. The data reported by Humphreys (1966) was not in a form which permitted comparison with the developed predictions of Nakayama (1968) and Majumdar et al (1976). Nevertheless an assessment of the influence of rotation on mean Nusselt number, Nu_m, is shown in figure 4.7. Here the rotational Rayleigh number, Ra_b, has been defined in terms of the mean difference between the wall and bulk temperature and further uses the diameter of the rube according to

$$Ra_b = \frac{\Omega^2 H \beta \rho^2 c_p \ d^3 (T_w - T_b)_L}{\mu k} \qquad (4.13)$$

It is interesting to note from figure 4.7 that the influence of centripetal buoyancy in this study is quite small and only really detectable at the lowest Reynolds number tested. Increased heat transfer with respect to increases in the rotational Reynolds number are, however, more marked. This was a consistent feature resulting from the work of Humphreys (1966).

Le Feuvre (1968) reported mean heat transfer data for air flowing in internal circular passages in a model rotor used to simulate the coolant channels in a drum-type electrical armature. Figure 4.8 shows the schematics of the apparatus used. The main rotor consisted of a cylindrical drum (1) fitted onto three discs (2) to which could be fitted either 8, 12 or 16 equispaced tubes (3) having a bore diameter of 19.1 mm and a nominal length of 304.8 mm. The eccentricity was either 50.8 mm or 76.2 mm. This rotor drum could rotate inside a fixed casing (4) made from perspex and arranged to have an inlet (5) and exit (6) chamber as shown in the figure. Baffles fitted into these chambers permitted adequate mixing of an air flow circulated through the tubes fitted to the rotor. One of these tubes was made from stainless steel and electrically heated using a wire wound heater. Ceramic sprayed onto the outer surface of this test tube permitted electrical insulation between the wire and tube surface and also good bonding. The tube wall temperature was measured at five equispaced locations using thermocouples spot welded onto the outer surface. The rotor could be driven at a fixed speed of 1500 rev/min by means

Reference	Fluid	L/d	H/d	Re	Ra_b	J_a	Pr
Humphreys (1966)	Air	48	48	$5 \times 10^3 - 2 \times 10^4$	$3 \times 10^4 - 1.8 \times 10^5$	5 - 15	0.7
		12	12	$5 \times 10^3 - 2 \times 10^4$	$4.8 \times 10^6 - 1.8 \times 10^7$	65 - 180	0.7
Humphreys, Morris and Barrow (1967)	Air	12	6	$5 \times 10^3 - 2 \times 10^4$	$4.8 \times 10^6 - 1.8 \times 10^7$	250 - 650	0.7
Le Feuvre (1968)	Air	10.7	1.3-2.0	$5 \times 10^3 - 4 \times 10^4$	$4.4 \times 10^6 - 6.7 \times 10^6$	480	0.7
Morris and Woods (1978)	Air	34.7	24	$5 \times 10^3 - 2 \times 10^4$	$2 \times 10^3 - 3.3 \times 10^5$	15 - 150	0.7
		69.3	48	$5 \times 10^3 - 2 \times 10^4$	$6 \times 10^2 - 2.2 \times 10^4$	6 - 40	0.7

TABLE 4.3   RANGE OF EXPERIMENTAL VARIABLES AVAILABLE FOR DEVELOPING TURBULENT FLOW.

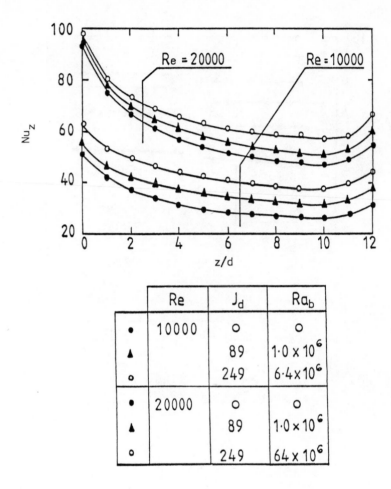

FIG. 4.6   EFFECT OF ROTATION ON DEVELOPING TURBULENT
HEAT TRANSFER.   HUMPHREYS, MORRIS AND
BARROW (1967).

126

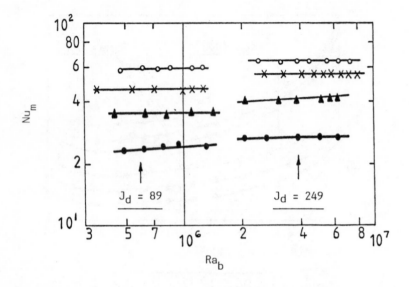

FIG. 4.7   EFFECT OF ROTATION ON TURBULENT HEAT
TRANSFER. HUMPHREYS, MORRIS AND
BARROW (1967) (L/d = 12, H/d = 6).

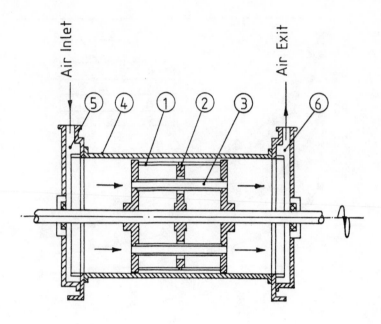

FIG. 4.8   SCHEMATIC ARRANGEMENT OF APPARATUS USED BY
            LE FEUVRE (1968).

of an electric motor directly coupled to the rotor.   Table 4.3 shows
the range of variables studied.

The experimental data of Le Feuvre (1968) was presented in a form
much more suitable to electrical machine designers but the results
did show broad quantative agreement with those presented by Humphreys
et al (1967).   A more detailed comparison of the various results for
mean heat transfer in the entry region will be given later in this
section.

Morris and Woods (1978) attempted to account for the improved heat
transfer due to rotation in the entry region by correlating mean heat
transfer for the experimental data of Woods (1975) with the rotational
Reynolds number, J_a.   Table 4.3 shows the range of variables studied.

The arguments given for the method of data presentation may be
briefly outlined as follows: We have seen from the work of Humphreys
et al (1967), Le Feuvre (1968) and the developed heat transfer data
of Woods (1975) shown in figure 4.3 that correlation of air data in
terms of centripetal buoyancy at Reynolds number values greater than
about 10,000 was not good.   This, in view of the comments made earlier
that Coriolis acceleration is important in regions of developing flow,
led them to disregard the rotational Rayleigh number as an important

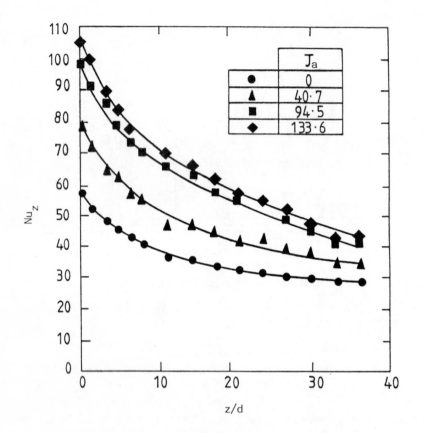

FIG. 4.9   TYPICAL EFFECT OF ROTATION ON DEVELOPING
TURBULENT HEAT TRANSFER. MORRIS AND WOODS
(1978) (L/d = 34.7, H/d = 24).

parameter and adopt the rotational Reynolds number as the sole charac-
terisation of rotational effects. Note that although this was also
done for their data with laminar flow, chronologically the turbulent
data was in fact first treated in this manner. Also Humphreys et al
(1967) and Le Feuvre (1968) had mentioned the influence of so-called
entry swirl at the entrance to their test sections due to upstream
plumbing arrangements being influenced by rotation. This is, in
effect, a Coriolis effect which further justifies the use of the rotational Reynolds number.

Figure 4.9 shows the typical way in which rotation was found to influence the axial distribution of local Nusselt number, Nu_z. The
trends are very similar to those detected by Humphreys et al (1967)
and clearly show the improved heat transfer with increases in rotational speed.

For an effective length/diameter ratio of 34.65 and an eccentricity parameter of 24, figure 4.10 shows the improvement in mean Nusselt number for two typical Reynolds number values and the same geometry defined in figure 4.9. The data scatter present in these results at each nominal Reynolds number occurs because tests were conducted with a variety of heat flux levels which, because of property variations with temperature, make precise control of Reynolds number difficult. This figure, consequently, is shown to illustrate the general pattern of heat transfer response. A similar trend was found with a test section having a length/diameter ratio of 69.3 and an eccentricity parameter of 48.

Initially Morris and Woods (1978) attempted to correlate the effect of rotation by examining the difference in mean Nusselt number obtained with rotation to that at zero rotational speed and the same Reynolds number. This proved unsuccessful due to data scatter effects. They were, nevertheless, able to correlate their data for both rotating test geometries studied with a simple exponent-type relationship between the mean Nusselt number, the Reynolds number and the rotational Reynolds number. The actual equation proposed which correlated data for both tubes to within ±12% was

$$Nu_m = 0.015 \ Re^{0.78} \ J_a^{0.25} \qquad\qquad (4.14)$$

Figure 4.11 shows the comparison of all test data for both geometries with equation (4.14). It is evident that the data obtained by these workers appear to be independent of the length/diameter ratio and the eccentricity parameter.

It is interesting now to conclude this section with an attempt to relate the data reported by Humphreys (1966), Humphreys et al (1967) and Le Feuvre (1968) with the results of Morris and Woods (1978). Although it is not possible to precisely pin-point each data point from the works of Humphreys (1966), Humphreys et al (1967) and Le Feuvre (1968) the cross hatched regions shown in figure 4.12 approximately span their range of validity when plotted in a manner suggested by equation (4.14).

Within the limits of experimental scatter the data of Humphreys (1966), Humphreys et al (1967) and Le Feuvre (1968) may be correlated by the equation

$$Nu_m = 0.012 \ Re^{0.78} \ J_a^{0.18} \qquad\qquad (4.15)$$

Although equation (4.15) has similar characteristics to that proposed by Morris and Woods (1978) it does suggest noticeably less enhancement in heat transfer. This is probably due to different velocity profiles at the immediate entry planes of the heated test sections reflecting differences in upstream delivery plumbing as described above.

This effect of upstream flow conditioning has not been systematically studied but Humphreys (1966) reports the results of two comparative tests made with deliberate attempts to demonstrate this effect. Figure 4.13 shows the effect of fitting flow straighteners, having

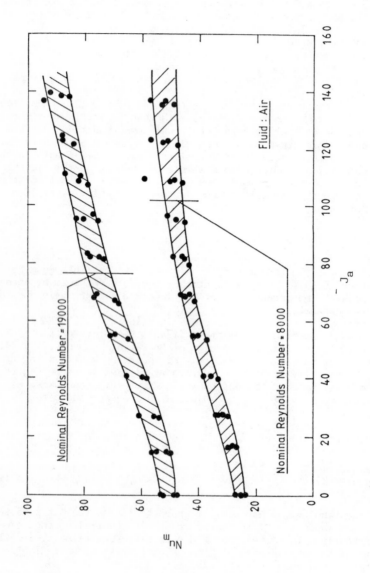

Nominal Reynolds Number = 19000

Nominal Reynolds Number = 8000

Fluid : Air

Nu_m

$- J_a$

FIG. 4.10   TYPICAL EFFECT OF ROTATION ON TURBULENT MEAN HEAT TRANSFER.   MORRIS AND
WOODS (1978) (L/d = 347, H/d = 24).

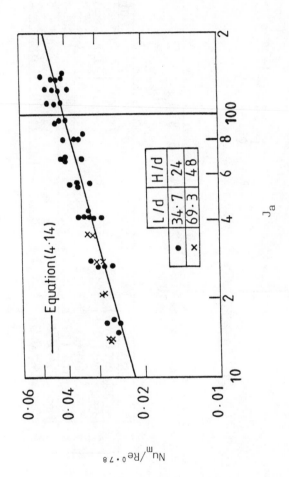

FIG. 4.11   CORRELATION OF MEAN TURBULENT HEAT TRANSFER PROPOSED BY MORRIS AND WOODS (1978).

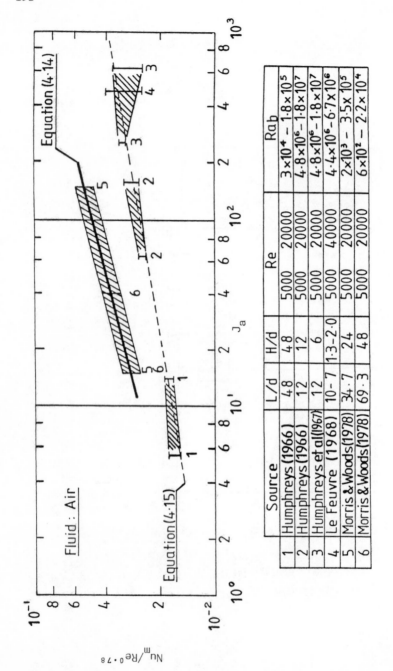

FIG. 4.12 COMPARISON OF ALL KNOWN DATA FOR MEAN TURBULENT ENTRY REGION HEAT TRANSFER WITH PROPOSAL OF MORRIS AND WOODS (1978).

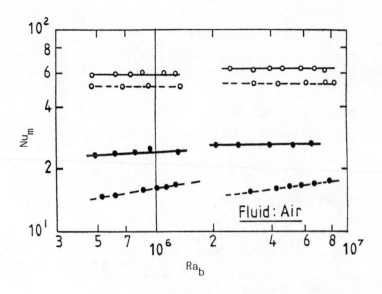

	Nominal Reynolds Number
●	5000
○	20000

——————— Flow straighteners not fitted in entry

— — — — — Flow straighteners fitted in entry

FIG. 4.13   EFFECT OF ENTRY PLANE FLOW CONDITIONING
ON HEAT TRANSFER PERFORMANCE. HUMPHREYS
(1966) (L/d = 12, H/d = 6).

the form of a tube bundle, fitted into the short calming length which preceded the entry plane of his test section. The level of heat transfer enhancement is significantly reduced when straighteners are fitted.

4.3 Recommendations for Design Purposes

The work reviewed in this chapter permits the following conclusions to be drawn which will be relevant to the design of cooling systems using this flow geometry and turbulent flow.

Flow Resistance with Unheated Flow

The only data available for the assessment of flow resistance with unheated turbulent flow is that reported by Morris (1981) and discussed in section 3.2 of the last chapter. The recommendations given in that section must consequentially also apply for turbulent flow.

Flow Resistance in Heat Flow

The only data pertaining to flow resistance with heated flow which can conveniently be utilised for design-type calculations is that due to Nakayama (1968) for developed flow. Thus equations (4.7) and (4.9) are recommended as a method for estimating pressure loss for circular tubes with respectively gas and liquid flows and fairly large length/diameter ratios. Note that the possible delaying of transition to turbulent flow as a result of the stabilising influence of rotation has not been investigated for heated flows to date.

Heat Transfer in Developed Flow

Theoretical studies of developed turbulent heat transfer suggest that centripetal buoyancy effects will not be as pronounced as with the case of laminar flow and the influence of Coriolis acceleration is negligible. For gas flows the predictions of Nakayama (1968) and Majumdar et al (1977) are in reasonable mutual agreement but the only currently available experimental data (obtained with test sections having L/d ratios in the range 34.7 - 69.3) tends to give heat transfer enhancement significantly higher than that predicted. This is probably due to the complex interaction of Coriolis acceleration and centripetal acceleration effects in the still developing region of the tube. It is suggested that the heat transfer enhancement for gas flows is estimated using equation (4.1) for locations at least 100 effective diameters downstream of the entrance plane. For length/diameter ratios less than 100 equation (4.1) will tend to be conservative in its estimation of heat transfer enhancement.

It appears from the experimental data of Nakayama and Fuzioka (1978) for developed water flows that the proposal of Nakayama (1968) for liquids gives tolerably good predictions. In view of this limited data, where seemingly the influence of Coriolis acceleration is not as marked, it is suggested that equation (4.4) may be used to estimate heat transfer enhancement with liquid flows having Prandtl number values greater than unity.

No information is currently available with which to determine the
entry length required to establish fully developed conditions with
heated turbulent flow of either gases or liquids.

Heat Transfer in the Entrance Region

Theoretical predictions of turbulent heat transfer in the entrance
region have not been fully explored to date and only air flows have
been experimentally studied. It is apparent that the Coriolis acceleration affects the problem markedly in the immediate entrance region
so that, for tubes having a length/diameter ratio upto about 100 and
gas-like flows, the heat transfer enhancement may be expressed in
terms of the rotational Reynolds number.

Equation (4.14) proposed by Morris and Woods (1978) is a simple
equation with which to estimate mean heat transfer enhancement for
gas flows in the range useful for turbo-generator rotor cooling systems. However based on an assessment of data from Humphreys (1966)
and Le Feuvre (1968) equation (4.15) tends to overestimate the enhancement. Consequently equation (4.15) may also be used for gases and
a compromise estimate made in relation to the result obtained from
using equation (4.14). The data used for the construction of equation
(4.14) and (4.15) was obtained with test sections not arranged to
smooth out upstream flow disturbances. When a deliberate attempt is
made to calm or straighten the flow prior to the commencement of heating the enhancement in mean heat transfer due to rotation is likely
to be lower than that suggested by equations (4.14) and (4.15). More
precise recommendations cannot be made in view of the current state
of knowledge and entry region data for liquid flows is not available.

CHAPTER 5

FLOW AND HEAT TRANSFER IN SQUARE-SECTIONED

TUBES WHICH ROTATE ABOUT A PARALLEL AXIS

WITH LAMINAR OR TURBULENT FLOW

5.1 Introduction

When practical examples of rotating cooling systems were cited in Chapter 1 it was pointed out that ducts of non-circular cross section would often be encountered. The internal cooling of turbo-generator rotor conductors is a case in point where, in the parallel mode of rotation, rectangular ducts are a common feature of modern designs.

As far as the present author is aware the only non-circular cross section which has been studied in the parallel mode is that having a square cross section. This chapter discusses this particular rotating geometry and, because the extent of the reported information is not as comprehensive as its circular cross sectioned counterpart, the case of laminar and turbulent flow will be presented in the one chapter. In principle all the arguments discussed in Chapters 3 and 4 are equally applicable since only the duct shape has changed; the underlying physical principles being identical.

Apart from reiterating the fact that rotation does not affect laminar flow resistance in this case once established flow exists, there is no additional information directly available for the estimation of isothermal flow resistance. This chapter is consequently devoted to reviewing the limited data available for heated flows.

5.2 Laminar Heated Flow in Square-Sectioned Ducts

5.2.1 Theoretical Studies of Developed Flow

Dias (1978) and Morris and Dias (1981) theoretically studied the case of heated developed laminar flow. With the exception that the duct was now square in section the problem was posed in exactly the same manner as the corresponding circular-tube problem reported by Woods and Morris (1974). The only main tactical difference between the two studies was that Morris and Dias (1981) followed the arguments presented in Chapter 2 more rigorously so that the interaction of a temperature-dependent density with the Coriolis acceleration was not included in their model. This effectively removed the influence of Coriolis acceleration explicitly from their analysis. Note that all the works reported on the circular tube with developed flow, where

138

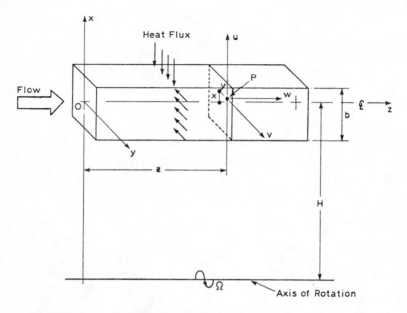

FIG. 5.1   COORDINATE SYSTEM FOR THE ROTATING SQUARE
SECTIONED TUBE.

density variations in the Coriolis terms had been included, demonstrated a very weak Coriolis effect in the parallel mode of rotation.

Figure 5.1 shows the flow system where the motion is referred to a
Cartesian frame of reference with flow mainly along the z-direction.
The conservation equations for stream function, vorticity, axial vel-
ocity and temperature may be manipulated as before to give

$$\nabla^2 \psi = - \xi \tag{5.1}$$

$$\nabla^2 \xi + \frac{\partial(\psi, \xi)}{\partial(X, Y)} - Ra_\tau \left[(\frac{X}{\epsilon_b} + 1) \frac{\partial \eta}{\partial Y} - \frac{Y}{\epsilon_b} \frac{\partial \eta}{\partial X} \right] = 0 \tag{5.2}$$

$$\nabla^2 W + \frac{\partial(\psi, W)}{\partial(X, Y)} + Re_p = 0 \tag{5.3}$$

$$\nabla^2 \eta + Pr \frac{\partial(\psi, \eta)}{\partial(X, Y)} + W = 0 \tag{5.4}$$

where the required acceleration terms are taken from equation (2.15).

In this case the independent and dependent variables have been non-
dimensionalised according to

$$X = \frac{x}{b} \ , \ Y = \frac{y}{b} \tag{5.5}$$

$$W = \frac{wb}{\upsilon} \ , \quad = \frac{(T_w - T)}{\tau b \, Pr} \tag{5.6}$$

and the non-dimensional groups which emerge are

$$Ra_\tau = \frac{\Omega^2 H \beta \, \tau b^4}{\alpha \upsilon^2} \qquad \text{(Rotational Rayleigh Number)}$$

$$Pr = \frac{\upsilon}{\alpha} \qquad \text{(Prandtl Number)}$$

$$Re_p = \frac{-b}{\rho \upsilon^2} \frac{\partial P'}{\partial z} \qquad \text{(Pseudo Reynolds Number)}$$

$$\varepsilon_b = \frac{H}{b} \qquad \text{(Eccentricity Parameter)}$$

$$\tag{5.7}$$

Note that all main symbols are the same as those used in the case of the circular tube and that the width of the square cross section, b, is used instead of the tube radius in the definition of parameters given in equation (5.7). Thus in this case the characteristic length is the hydraulic diameter of the duct. Also the non-dimensional stream function is linked to the cross stream velocity components according to

$$u = \upsilon \frac{\partial \Psi}{\partial y} \ , \ v = - \upsilon \frac{\partial \Psi}{\partial x} \tag{5.8}$$

Equations (5.1) through (5.4), which must now satisfy the boundary conditions

$$W = \frac{\partial \Psi}{\partial x} = \frac{\partial \Psi}{\partial y} = \eta = 0 \text{ at } X = \pm \frac{1}{2} \text{ and } Y = \pm \frac{1}{2} \tag{5.9}$$

were solved numerically using the computational method proposed by Gosman et al (1968). The vorticity boundary conditions at the wall were again approximated using truncated versions of the full partial differential equations controlling the problem and full details are available in Dias (1978).

Figure 5.2 shows for a range of Prandtl number values the enhancement in heat transfer brought about by rotation resulting from the numerical solution. The enhancement is expressed relative to the stationary condition which in thise case gives a developed bulk Nusselt number of 3.61 (see for example Morris (1968)).

Figure 5.3 shows the corresponding impediment to flow which results from the creation of the buoyancy-induced secondary flow. Figures

140

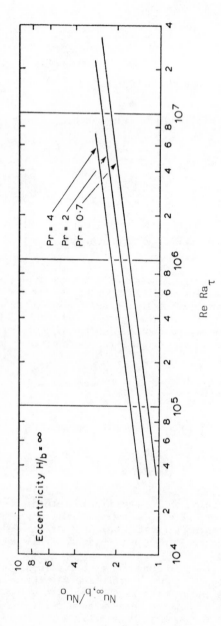

FIG. 5.2 EFFECT OF ROTATION ON DEVELOPED LAMINAR HEAT TRANSFER IN SQUARE-SECTIONED TUBES. (MORRIS AND DIAS (1981).

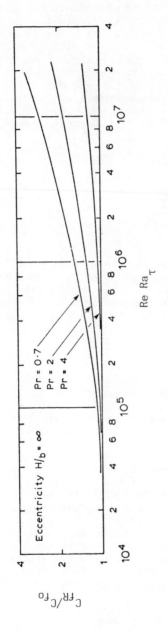

FIG. 5.3   EFFECT OF ROTATION ON DEVELOPED LAMINAR FLOW RESISTANCE WITH HEATED FLOW IN A
SQUARE-SECTIONED TUBE.   MORRIS AND DIAS (1981).

5.2 and 5.3 both refer to an eccentricity $\varepsilon_b = \infty$. Numerical experimentation with the solution procedure demonstrated little sensitivity to eccentricity. The trends noted for the square-sectioned tube were identical with those reported by Woods and Morris (1978) for the circular-sectioned counterpart and need no further comment. For completeness figure 5.4 shows the heat transfer enhancement plotted according to the implication of the high Prandtl number asymptote and figure 5.5 shows typical distributions of the dependent variables. The high Prandtl number asymptote was again found to be useful even at Prandtl number values in the gas-like flow ranges. Over the range $4 \times 10^4 \leqslant$ Re Ra$_\tau$ Pr $\leqslant 4 \times 10^6$ the theoretically determined enhancement in heat transfer was well approximated by

$$\frac{Nu_{\infty,b}}{\overline{Nu}_0} = 0.21(\text{Re Ra}_\tau \text{ Pr})^{0.16} \qquad (5.10)$$

5.2.2 Experimental Studies of Laminar Developed Flow and Comparison With Theoretical Predictions.

Dias (1978) and Morris and Dias (1981) also conducted an experimental study of developed laminar heat transfer in this flow configuration using air as the test fluid. They modified the apparatus used by Woods and Morris (1974) to incorporate a heated square-sectioned duct which was 610 mm long by 9.53 mm square internally. This test section could be rotated with a centre-line eccentricity of either 304.8 mm or 457.2 mm. In principle the same experimental procedure and data processing method as that used by Woods and Morris (1974) was adopted.

Figure 5.6 shows a comparison of the experimentally determined enhancement in heat transfer in relation to the theoretically predicted valued for both eccentricity levels tested. The developed values of Nusselt number used in this comparison were calculated from the average of five axial locations, at which temperature measurements were made, in what was deemed to be the closest approximation to developed flow. This tended to cover the axial location range $37 \leqslant z/b \leqslant 51$ along the test section which had an overall length/hydraulic diameter ratio of 64. The zero speed reference condition with which to measure heat transfer enhancement was taken from experimental data taken with the same apparatus when stationary. The following observations may be made on reference to figure 5.6.

Although the numerical solution over-predicts the test data, as was the case with the circular tube, the trend of the results show reasonable agreement with that predicted. There was evidence that data obtained at the higher eccentricity ratio was in closer agreement with the predicted behaviour. However, on average, a smoothing curve through the experimental data is typically 25% below that predicted theoretically.

It may be argued that the data scatter shown for Re Ra$_\tau$ > 10^7 could be interpreted as a progressive reduction in heat transfer enhancement

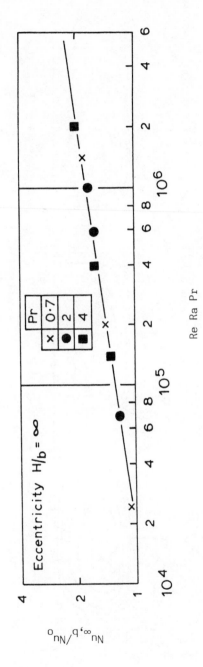

FIG. 5.4 COMPARISON OF THE HIGH PRANDTL NUMBER ASYMPTOTE WITH PREDICTIONS AT LOWER
PRANDTL NUMBER VALUES. MORRIS AND DIAS (1981).

144

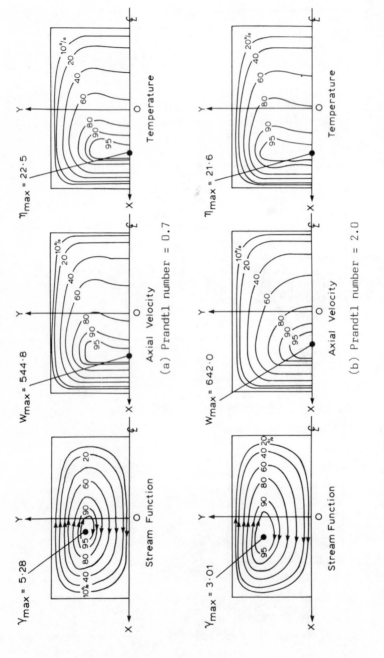

FIG. 5.5   TYPICAL PREDICTIONS OF STREAM FUNCTION, AXIAL VELOCITY AND TEMPERATURE FOR LAMINAR DEVELOPED FLOW IN A ROTATING SQUARE-SECTIONED TUBE.   DIAS (1978) AND MORRIS AND DIAS (1981).

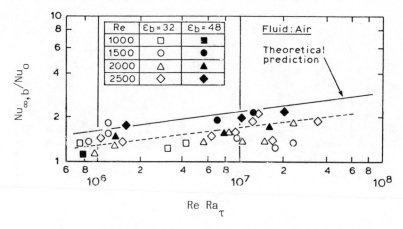

Re Ra$_\tau$

FIG. 5.6 COMPARISON OF EXPERIMENTAL DATA FOR DEVELOPED
LAMINAR HEAT TRANSFER WITH THEORETICAL PREDICTIONS.
DIAS (1978) AND MORRIS AND DIAS (1981).

as noted with air and water in the case of the circular-sectioned
tube, but the effect is not conclusive.

5.2.3 Laminar Heat Transfer in the Entrance Region

Theoretical predictions of laminar developing heat transfer have
not, to date, been published although the work reported by Majumdar
et al (1977) for turbulent flow is capable of being adapted to conditions of laminar flow. However, in their experimental studied of developed laminar heat transfer Dias (1978) and Morris and Dias (1981)
have made experimental observations in the developing region of a
square-sectioned duct and their main findings will now be discussed.

Figure 5.7 shows the typical enhancement in local Nusselt number
which was found in this instance. The data in this figure was obtained
with a nominal Reynolds number of 2000 and an eccentricity ratio of
32. As the rotational speed, expressed as a rotational Reynolds number, J_b, using the hydraulic diameter as a length characteristic, increases there is a systematic increase in local heat transfer again
as noted with circular tubes.

(Note in this work that $J_b = \dfrac{\Omega^2 b^2}{\upsilon}$).

Figure 5.8 illustrates the effect of heat flux at a fixed level of
rotational Reynolds number. As the heat flux increases the local heat
transfer also increases. This reflects the complex interaction between the Coriolis acceleration effect in the entry region with the
centripetal buoyancy.

146

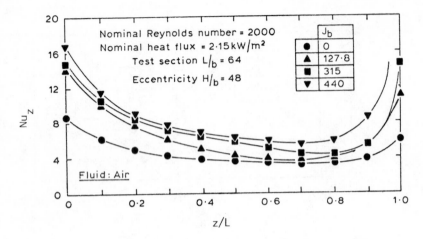

FIG. 5.7   TYPICAL EFFECT OF ROTATION ON LOCAL LAMINAR
            NUSSELT NUMBER FOR A SQUARE-SECTIONED TUBE.
            DIAS (1978) AND MORRIS AND DIAS (1981).

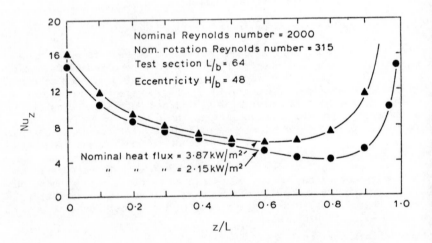

FIG. 5.8   TYPICAL EFFECT OF HEAT FLUX ON LOCAL LAMINAR
            NUSSELT NUMBER FOR A ROTATING SQUARE-SECTIONED
            TUBE. DIAS (1978) AND MORRIS AND DIAS (1981).

At a fixed level of heat flux and rotational Reynolds number, figure 5.9 demonstrates the effect of eccentricity which was typically found. It is reasonable to assume that based on this information that eccentricity does not have a strong effect other than via the rotational Rayleigh number.

Being led by the previous work for circular tubes from Woods and Morris (1978), figure 5.10 shows the variation of the mean Nusselt number Nu_m, with rotational Reynolds number for two values of the through flow Reynolds number. The mean value was determined for a length/effective diameter ratio of 48. The same remarks apply concerning the data scatter as explained in Chapter 3 for the circular tube. Although there is a centripetal buoyancy effect in the entrance region see figure 5.8, a simple design-biased correlation of the mean heat transfer data was attempted using only the rotational Reynolds number as a representation of angular velocity.

Thus, for $120 < J_b < 620$, Morris and Dias (1981) proposed that mean Nusselt number could be estimated from

$$Nu_m = 0.011 \ Re^{0.78} \ J_b^{0.11} \tag{5.11}$$

Note that this data was obtained effectively with $L/b = 48$ and eccentricities of 32 and 48 respectively and Reynolds number values less than 2000. Figure 5.11 shows all data points in relation to equation (5.11). Clearly the precise influence of centripetal buoyancy is not being fully taken into account and this probably accounts for the data scatter evident in figure 5.11. Note that this form of correlation was originally used for turbulent flow in circular tubes in the first instance where buoyancy is relatively less dominant and its extension to laminar flow is only in the interest of simplicity. It is interesting to note that the exponent of the rotational Reynolds number is considerably smaller than that obtained with the circular tube, see comparison with equations (3.127 and (3.128). The use of circular tube correlations and a hydraulic diameter concept would seriously overpredict the heat transfer.

5.3 Turbulent Heated Flow in Square-Sectioned Ducts

Not a great deal of data is available from the technical press concerning turbulent flow with this particular rotating geometry. The theoretical investigation of Majumdar et al (1977) described in Chapter 4 included some aspects of flow and heat transfer in the square-sectioned tube but a systematic investigation of the controlling physical parameters was not given. The work demonstrates that the computational method may be applied to this problem nevertheless. Some agreement with experimentally measured developed Nusselt numbers was quoted but precise details were not available.

The work outlined earlier in this chapter by Dias (1978) was extended, see Morris and Dias (1981), to include an experimental study of turbulent flow with air and this work is the only currently available source with which to make design-type recommendations. The work was performed with the same apparatus and range of geometric variables as

148

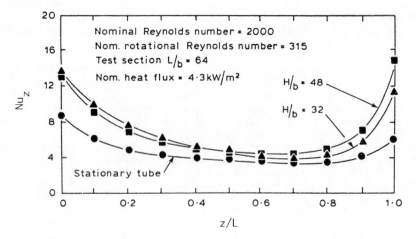

FIG. 5.9   TYPICAL EFFECT OF ECCENTRICITY ON LOCAL LAMINAR
NUSSELT NUMBER FOR A ROTATING SQUARE-SECTIONED
TUBE.   DIAS (1978) AND MORRIS AND DIAS (1981).

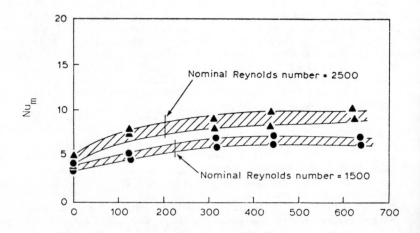

FIG. 5.10   TYPICAL EFFECT OF ROTATION ON MEAN LAMINAR
NUSSELT NUMBER FOR A ROTATING SQUARE-SECTIONED
TUBE.   DIAS (1978) AND MORRIS AND DIAS (1981).

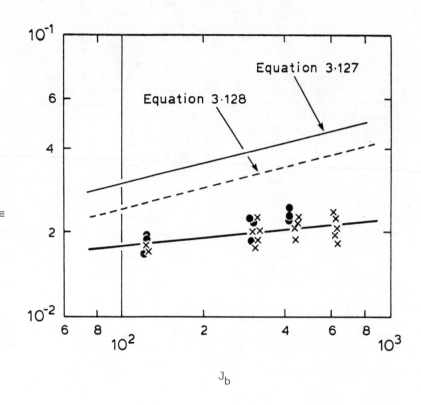

FIG. 5.11   PROPOSED CORRELATION OF MEAN LAMINAR
NUSSELT NUMBER FOR A ROTATIONAL SQUARE-
SECTIONED TUBE. DIAS (1978) AND
MORRIS AND DIAS (1981).

their study of laminar flow described in section 5.2. Their salient findings may be summarised as follows.

It was generally found that for a specified through flow Reynolds number and heat flux setting, the heat transfer tended to be improved with increases in the rotational speed in exactly the same qualitative manner as previously reported data taken with circular-sectioned tubes

Figure 5.12 typifies the manner in which the local Nusselt number responded to increases in rotational speed. Here, for the operating conditions shown in the figure, the local Nusselt number is plotted against the normalised axial location along the heated tube. At each rotational speed, the local Nusselt number distributions had similar features. Firstly the initially high region of heat transfer which asymptotes towards a plateau region consistent with the customary boundary layer development which occurs in a heated tube. Towards the exit end of the test section (notionally the final 20%) the up-swing in the local heat transfer usually associated with a combination of exit-type effects and end conduction losses were evident. The important feature to note is that there is a systematic improvement in the local heat transfer with increasing rotational speed. Note that the rotational Reynolds number, J_b, has been used to describe rotation since the high Reynolds number flows associated with turbulence buoyancy was felt to be the less dominating effect as described earlier.

The trends shown in figure 5.12 were obtained with tests undertaken with an eccentricity of 304.8 mm between the axis of rotation and the central axis of the tube, that is ε_b = 32. Analysis of data obtained with a repeated programme of experiments but with an eccentricity of 457.2 mm (ε_b = 48) produced identical results qualitatively but with a tendency for the heat transfer enhancement with rotation to be marginally higher at the higher value of eccentricity parameter. With a nominal through flow Reynolds number of 20000, figure 5.13 exemplifies the comparison of local Nusselt number variations for the two eccentricities examined.

Two possible explanations may be proposed for the eccentricity effect shown in figure 5.13. Firstly since the Coriolis acceleration interacts with a developing flow field to create vorticity relative to the tube, then this effect will be influenced by the velocity profile at the plane where heating commences. This entry plane velocity profile is itself influenced by upstream plumbing arrangements necessary to deliver the coolant to the test section. Thus the increased radial delivery tube required to accommodate the change of eccentricity also tends to increase the residual secondary flow present at the entry plane of the test section. This in turn effects the flow development and hence the heat transfer via the Coriolis interaction.

Secondly, for a given rotational speed, an increase in the eccentricity parameter implies an increase in the centre line centripetal acceleration. This in turn implies an increase in the relative strength of any buoyancy interaction and, since in this case buoyancy tends to improve heat transfer, it is likely that this effect also influences the local heat transfer response. Both these effects offer areas where an extension of the work currently being reported may be further developed in the future and also apply to the case of laminar flow.

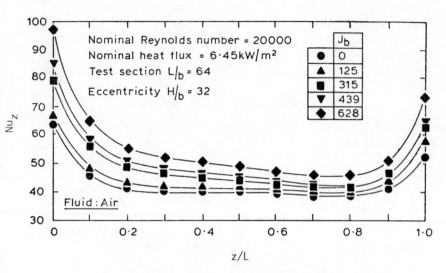

FIG. 5.12   TYPICAL EFFECT OF ROTATION ON LOCAL TURBULENT
NUSSELT NUMBER FOR A SQUARE-SECTIONED TUBE.
DIAS (1978) AND MORRIS AND DIAS (1981).

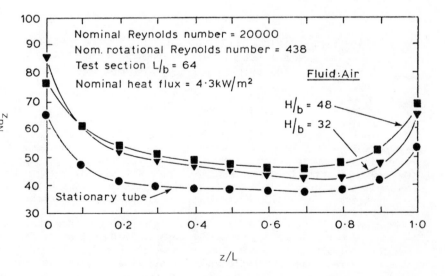

FIG. 5.13   TYPICAL EFFECT OF ECCENTRICITY ON LOCAL
TURBULENT NUSSELT NUMBER FOR A SQUARE-
SECTIONED TUBE. DIAS (1978) AND MORRIS
AND DIAS (1981).

152

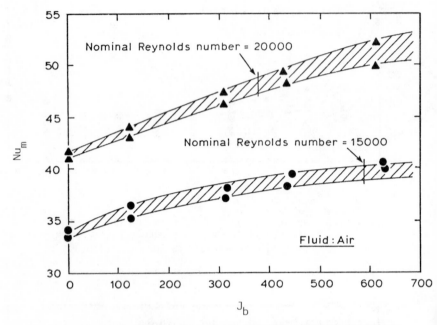

FIG. 5.14   TYPICAL EFFECT OF ROTATION ON MEAN TURBULENT
HEAT TRANSFER IN A SQUARE-SECTIONED TUBE.
DIAS (1978) AND MORRIS AND DIAS (1981).

The improvement in local heat transfer with increased rotational
speed is also reflected in the mean level of heat transfer for the
tube.   Thus figure 5.14 illustrates the effect of speed changes on the
mean Nusselt number which was typically measured.   The mean Nusselt
number is defined in terms of the mean heat flux for the test section
with the motivating temperature difference taken as that between the
mean wall temperature and mean fluid temperature between the inlet and
exit stations.   The following features should be noted in relation to
figure 5.14.

Although data is shown for two nominal values of the through flow
Reynolds number, the actual value for each individual test differs
from the nominal value quoted since changes in heat flux loads, system
temperature response to rotations, etc., all have an attendant influence on fluid properties which make exact control of the through flow
Reynolds number extremely difficult.   It is for this reason mainly
that the results shown have the data scatter band widths shown. There
was also a slight tendency for the Nusselt number to increase with
increases in heat flux (other parameters notionally fixed) which sug-

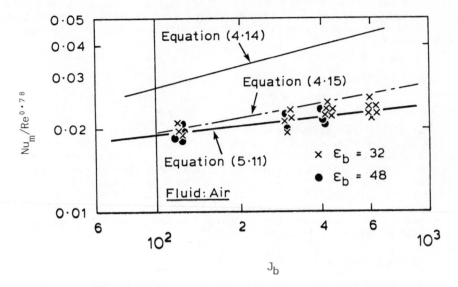

FIG. 5.15   OVERALL INFLUENCE OF ROTATION ON MEAN
TURBULENT HEAT TRANSFER IN SQUARE-SECTIONED
TUBE.   DIAS (1978) AND MORRIS AND DIAS (1981).

gested that a buoyancy effect was still present to some extent even in
the turbulent flow regime considered. Figure 5.14 however clearly
illustrates the significant improvement in mean Nusselt number which
results from increases in the rotational Reynolds number.

Being led by the earlier work of Morris and Woods (1978) and their
attempt to correlate mean Nusselt numbers in a circular-sectioned tube
with a simple exponent-type equation along the lines of equation (4.14)
Morris and Dias (1981) tried a similar approach with their square-sectioned tube data.

As a result of a detailed examination of all data points taken with
the square-sectioned tube the following correlating equation was adopted

$$Nu_m = 0.012 \, Re^{0.78} \, J_b^{0.1} \qquad (5.12)$$

This equation correlated data with a maximum scatter of ± 14% and was
derived from data covering the range $120 < J_b < 620$, $32 < \varepsilon_b < 48$,
$5000 < Re < 20,000$ for $L/d = 48$. Figure 5.15 compares all the data
points, with both eccentricity values studied, with equation (5.12).
Also shown for comparative purposes is the corresponding circular-tube
equations of Morris and Woods (1978) and that proposed in the present
work from a survey of the works of Humphreys (1966) and Le Feuvre
(1968). It is interesting to note that the exponent of the rotational
Reynolds number arising from equation (5.14) is in better agreement

with that arising from equation (4.15). Note that equations (4.14) and (4.15) have been presented in terms of a hydraulic diameter length characteristic in figure 5.15 to permit direct comparison with equation (5.12).

5.4 Recommendations for Design Purposes.

Flow Resistance with Unheated Flow

No direct theoretical or experimental work is available with which to determine flow resistance in square-sectioned tubes rotating in the parallel mode. For laminar developed flow the arguments presented in the present monograph suggest that rotation will not influence flow resistance. This however will not be true in relatively short tubes where Coriolis-generated secondary flow occurs. Data available for circular tubes, see equations (3.7) and (3.14) may be coupled to a hydraulic diameter concept as a rough estimate but the results should be treated with caution.

Flow Resistance with Heated Flow

Only the theoretical prediction of Dias (1978) and Morris and Dias (1981) for developed laminar flow is currently available. The results of these investigations will be useful for laminar flow in relatively long square-sectioned ducts for all gas-like and liquid-like flows. No information is available with heated turbulent flow.

Heat Transfer with Laminar Flow

The theoretical predictions of Dias (1978) and Morris and Dias (1981) for developed laminar flow may be conveniently summarised by equation (5.10) which may be used to estimate expected heat transfer enhancement over a wide range of Prandtl number values. The only experimental data available is for air flow and the experimentally determined heat transfer enhancement was found to be typically 25% lower than that suggested by equation (5.10).

In the entrance region equation (5.11) may be used to estimate the heat transfer enhancement on a mean basis. This equation was determined experimentally and does not take into account quantitatively the full interaction between Coriolis acceleration and centripetal buoyancy. Even so it is the only information currently available and should not be extrapolated too far beyond the range $120 < J_b < 620$, $32 < \epsilon_b < 48$, $L/d = 48$. The use of circular-tube correlations together with an effective diameter is not recommended in view of the wide difference between the exponents of the rotational Reynolds numbers in correlations having the form of equations (3.127) and (3.128).

Heat Transfer with Turbulent Flow

Theoretical investigations particularly those involving numerical methods of solution have not yet been fully exploited and the only experimentally based correlation available is that due to Morris and Dias (1981) as given by equation (5.12) for mean Nusselt number. Correlations based on data from circular-sectioned tubes rotating in the parallel mode are not recommended for the square-sectioned tube linked to the hydraulic diameter concept.

CHAPTER 6

FLOW AND HEAT TRANSFER IN CIRCULAR-SECTIONED

TUBES WHICH ROTATE ABOUT AN ORTHOGONAL AXIS WITH

LAMINAR OR TURBULENT FLOW

6.1 Introduction

Thus far in this monograph attention has been focussed exclusively on tubes which rotate in the so-called parallel-mode. We now turn our attention to a review of the corresponding problem when the ducts considered are rotating about an axis perpendicular to the duct, the so-called orthogonal-mode rotation as depicted in figure 6.1 for the case of a circular-sectioned tube. This flow geometry, as we have seen from earlier comments, is particularly important in the design of cooled turbine rotor blades. The chapter will be structured along the same basic lines as adopted for Chapters 3 through 5 in that laminar flow will be considered initially to be followed by turbulent flow and in both cases isothermal and heated flows will be treated.

6.2 Isothermal Laminar Flow in Circular-Sectioned Ducts

One of the earliest attempts to theoretically examine flow in the orthogonal-mode of rotation was made by Barua (1955) who treated isothermal developed laminar flow. The flow and coordinate system is shown in figure 6.1 where it is seen that the positive outward direction of the z-axis is measured along the central axis of the duct from an origin having distance H from the rotational axis. The modifications due to rotation which must be made to determine the acceleration vector of a typical fluid particle were treated in Chapter 2 so that equation (2.29) is still applicable. Noting that the centripetal-type terms may be described using a scalar function, ϕ, permits the acceleration vector, $\underline{f}$ to be written as

$$
\begin{aligned}
\underline{f} = \underline{i}&\left[\frac{Du}{Dt} - \frac{v^2}{r} - 2\Omega w \sin\theta + \frac{\partial\phi}{\partial r}\right] \\
+ \underline{j}&\left[\frac{Dv}{Dt} + \frac{uv}{r} - 2\Omega w \cos\theta + \frac{1}{r}\frac{\partial\phi}{\partial r}\right] \\
+ \underline{k}&\left[\frac{Dw}{Dt} + 2\Omega(u\sin\theta + v\cos\theta) + \frac{\partial\phi}{\partial z}\right]
\end{aligned}
\tag{6.1}
$$

158

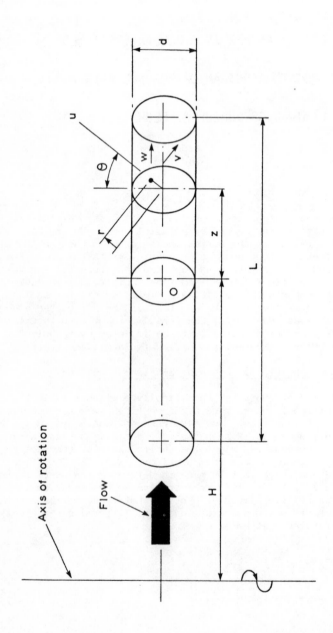

FIG. 6.1   ROTATING GEOMETRY AND COORDINATE SYSTEMS.

where u, v and w are the velocity components in the r, θ and z directions respectively, $\underline{i}$, $\underline{j}$ and $\underline{k}$ are the corresponding unit vectors and

$$\phi = \frac{1}{2}\left[\Omega^2 r^2 \sin^2 \theta + (H + z)^2\right] \tag{6.2}$$

When equation (6.1) is substituted into the Navier-Stokes equation for laminar constant property flow (see equation (2.37)) it is possible to combine the terms involving ϕ with those involving pressure, p, to give

$$\frac{Du}{Dt} - \frac{v^2}{r} - 2\Omega w \sin \theta = -\frac{1}{\rho}\frac{\partial \chi}{\partial r} + \upsilon\left[\nabla^2 u - \frac{u}{r^2} - \frac{2}{r^2}\frac{\partial v}{\partial \theta}\right] \tag{6.3}$$

$$\frac{Dv}{Dt} + \frac{uv}{r} - 2\Omega w \cos \theta = -\frac{1}{\rho r}\frac{\partial \chi}{\partial \theta} + \upsilon\left[\nabla^2 v - \frac{v}{r^2} - \frac{2}{r^2}\frac{\partial u}{\partial \theta}\right] \tag{6.4}$$

$$\frac{Dw}{Dt} + 2\Omega(u \sin \theta + v \cos \theta) = -\frac{1}{\rho}\frac{\partial \chi}{\partial z} + \upsilon\nabla^2 w \tag{6.5}$$

where

$$\chi = p - \rho\phi \tag{6.6}$$

If the flow is assumed to be established, so that axial gradients of velocity vanish, then continuity may be satisfied by means of a two-dimensional stream function, Ψ, defined as

$$\frac{\partial \Psi}{\partial r} = -\frac{v}{\upsilon} \quad , \quad \frac{\partial \Psi}{\partial \theta} = \frac{ru}{\upsilon} \tag{6.7}$$

The established flow assumption further implies that

$$\chi = \gamma z + G(r,\theta) \tag{6.8}$$

where γ is a constant axial pressure-type gradient and G is functionally related to cross stream variables alone.

Instead of using the vorticity transport equation Barua (1955) eliminated χ from equations (6.3) and (6.4) by cross differentiation and subtraction. When this is done the introduction of equations (6.7) and (6.8) together with the non-dimensional transformations

$$W = \frac{wa}{\upsilon} \qquad R = \frac{r}{a} \tag{6.9}$$

yields

$$\nabla^2\Psi + \frac{1}{R}\frac{\partial(\Psi,\nabla^2\Psi)}{\partial(R,\theta)} - 2J_a\left[\frac{\partial W}{\partial R}\cos \theta - \frac{1}{R}\frac{\partial W}{\partial \theta}\sin \theta\right] = 0 \tag{6.10}$$

where

$$J_a = \frac{\Omega a^2}{\upsilon} \qquad \text{(Rotational Reynolds number)} \qquad (6.11)$$

The introduction of equations (6.7), (6.8) and (6.9) into equation (6.5) similarly yields

$$\nabla^2 W + \frac{1}{R} \frac{\partial(\Psi,W)}{\partial(R,\theta)} + 4 \, Re_p + 2 \, J_a \left[\frac{\partial \Psi}{\partial R} \cos\theta - \frac{1}{R} \frac{\partial \Psi}{\partial \theta} \sin\theta \right] = 0 \qquad (6.12)$$

where

$$Re_p = \frac{-a^3}{4\rho\upsilon^2} \frac{\partial X}{\partial z} \qquad \text{(pseudo Reynolds number)} \qquad (6.13)$$

Barua (1955) derived an approximate solution for equations (6.10) and (6.12) by assuming that Ψ and W could be expanded in terms of the rotational Reynolds number J_a, as

$$\left. \begin{aligned} \Psi &= \Psi_0 + J_a \, \Psi_1 + J_a^2 \, \Psi_2 + \text{---} \\[2mm] W &= W_0 + J_a \, W_1 + J_a^2 \, W_2 + \text{---} \end{aligned} \right] \qquad (6.14)$$

Substitution of equations (6.14) into equations (6.10) and (6.12) permits sequential solutions for (Ψ_0,W_0), (Ψ_1,W_1), etc. to be determined and the solutions upto first order were found to be

$$\left. \begin{aligned} \Psi_0 &= 0 \\[2mm] W_0 &= Re_p(1-R^2) \end{aligned} \right] \qquad (6.15)$$

$$\left. \begin{aligned} \Psi_1 &= Re_p \, R(1-R^2)^2 \cos\theta/96 \\[2mm] W_1 &= -Re_p^2 \, R(1-R^2)(R^4 -3R^2+3) \sin \ /2304 \end{aligned} \right] \qquad (6.16)$$

To determine the influence of rotation on the mean flow through the pipe it is necessary to solve equation (6.14) upto second order terms. Barua (1955) actually accomplished this and proceeded to evaluate, in effect, the mean velocity in the axial direction. The mean velocity may be directly related to the usual through flow Reynolds number, Re, and, in terms of the nomenclature used in this monograph this leads to

$$Re = Re_p \left[1 - J_a^2 \left[\frac{1}{576} + \frac{Re_p^2}{1032192} \right] \right] \qquad (6.17)$$

If the resistance offered to flow is quantified in terms of the Blasius friction factor, C_{fR}, defined by equation (3.37), but using

$\frac{\partial X}{\partial z}$ as the pressure-type gradient, then it is easy to show that the ratio of friction factor under conditions of rotation with respect to the stationary case is simply

$$\frac{C_{fR}}{C_{fo}} = \frac{Re_p}{Re} \qquad (6.18)$$

where C_{fo} is the friction factor with a stationary tube.

Hence for the present analysis

$$\frac{C_{fR}}{C_{fo}} = \frac{1}{\left(1 - J_a^2\left[\frac{1}{576} + \frac{Re_p^2}{1032192}\right]\right)} \qquad (6.19)$$

Figure 6.2 illustrates the main features of the secondary flow pattern established in the R - θ plane as a consequence of rotation. Based on the solutions upto first order we note that the superposition of the axial velocity and the secondary flow shown in figure 6.2 results in a spiralling flow along the tube. The secondary flow is symmetrical about the diameter located at θ = 90° so that fluid particles in the plane passing through this diameter tend to remain in that plane. Also from equations (6.15) and (6.16) it is easy to show that the tangential velocity component vanishes identically at R = $1/\sqrt{5}$ for all angular locations. Further since the radial velocity component vanishes for all radial locations at θ = 0 or π then the stream lines for relative motion through the points R = $1/\sqrt{5}$, θ = 0 and R = $1/\sqrt{5}$, θ = π are straight lines. The motion may be considered as a screw-type motion with respect to these two lines.

The way in which rotation affects the axial velocity field is demonstrated typically in figure 6.3. The symmetrical parabolic profile when the tube is stationary distorts as shown with the level of distortion increasing as the rotation increases. The consequential influence of rotation on flow resistance is shown in figure 6.4. For a specified value of Re_p, which is tantamount to specifying the axial pressure gradient applied, the resistance progressively increases as rotation is increased. An alternative representation is given in figure 6.5 where the way in which the through flow Reynolds number responds to rotation is shown. It is clear that significant reduction in flow may occur as rotation increases.

Benton and Boyer (1966), Mori and Nakayama (1968) and Ito and Nanbu (1971) have attempted theoretical analyses of developed laminar flow by assuming that the flow may be treated as a central core region together with a relatively thin boundary layer in the immediate vicinity of the wall. This approach will be discussed in general terms firstly and then the particular results from these studies described in more detail.

Equations (6.3) through (6.5) may alternatively be non-dimensionalised using the transformations

162

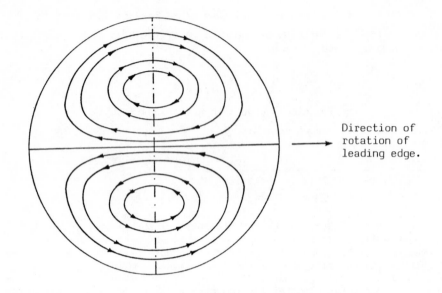

FIG. 6.2   TYPICAL SECONDARY FLOW PATTERN FOR
ISOTHERMAL LAMINAR DEVELOPED FLOW WITH
ORTHOGONAL-MODE PATTERN ROTATION FROM
BARUA (1955).

$$U = \frac{u}{w_m} \quad , \quad V = \frac{v}{w_m} \quad , \quad W = \frac{w}{w_m} \quad , \quad \chi^* = \frac{\chi}{pw_m{}^2}$$

$$R = \frac{r}{a} \quad , \quad Z = \frac{z}{a} \tag{6.20}$$

where w_m is the mean axial velocity, to give, for fully developed flow

$$U \frac{\partial u}{\partial R} + \frac{v}{R} \frac{\partial u}{\partial \theta} - \frac{V^2}{R} - \frac{J_d}{Re} W \sin \theta = - \frac{\partial \chi^*}{\partial R} + \frac{2}{Re}\left[\nabla^2 U - \frac{U}{R^2} - \frac{2}{R^2} \frac{\partial V}{\partial \theta}\right] \tag{6.21}$$

$$U \frac{\partial V}{\partial R} + \frac{V}{R} \frac{\partial V}{\partial \theta} + \frac{UV}{R} - \frac{J_d}{Re} W \cos \theta = - \frac{1}{R} \frac{\partial \chi^*}{\partial \theta} + \frac{2}{Re}\left[\nabla^2 V - \frac{V}{R^2} + \frac{2}{R^2} \frac{\partial U}{\partial \theta}\right] \tag{6.22}$$

$$U \frac{\partial W}{\partial R} + \frac{V}{R} \frac{\partial W}{\partial \theta} + \frac{J_d}{Re}\left[U \sin \theta + V \cos \theta\right] = - \frac{\partial \chi^*}{\partial Z} + \frac{2}{Re} \nabla^2 W \tag{6.23}$$

where
$$Re = \frac{w_m d}{\upsilon} \quad , \quad J_d = \frac{\Omega d^2}{\upsilon} \tag{6.24}$$

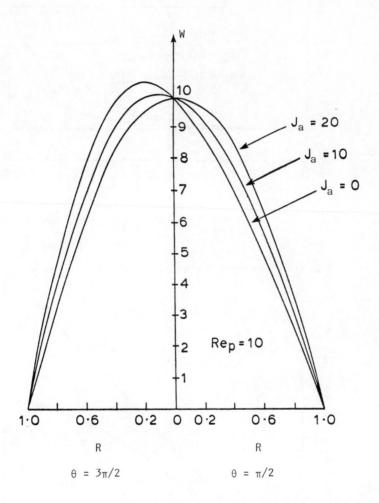

FIG. 6.3   TYPICAL EFFECT OF ROTATION ON AXIAL
VELOCITY PROFILES UPTO FIRST ORDER.
BARUA (1955).

164

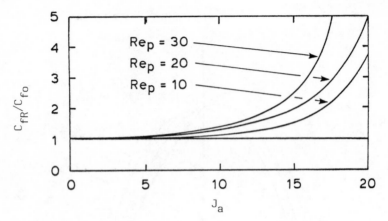

FIG. 6.4   INFLUENCE OF ROTATION ON FLOW RESISTANCE.
BARUA (1955).

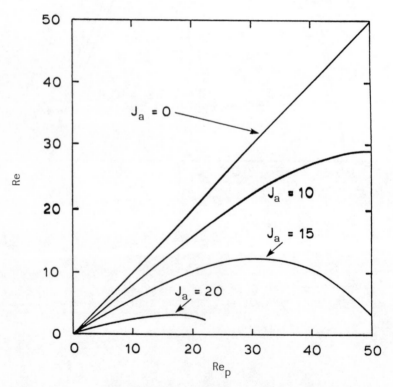

FIG. 6.5   INFLUENCE OF ROTATION ON MEAN FLOW RATE.
BARUA (1955).

Note that for convenience of later comparison with the experimental work presented by Ito and Nanbu (1971), the rotational Reynolds number J_d, has been defined using the tube diameter, d as the characteristic length dimension. Suppose for the present that developed laminar flow is considered for the case where Re $\gg J_d$ which implies low rotational speed in relation to the mean flow along the duct. The entire flow regime is considered to be made up of an inviscid core where the flow is mainly controlled by inertial and pressure forces alone together with a boundary layer region near the wall. If the subscript 'c' is used to denote independent variables in the core region then, for the core, we have

$$U_c \frac{\partial U_c}{\partial R} + \frac{V_c}{R} \frac{\partial U_c}{\partial \theta} - \frac{V_c^2}{R} - \frac{J_d}{Re} W_c \sin \theta = - \frac{\partial X^*_c}{\partial R} \qquad (6.25)$$

$$U_c \frac{\partial V_c}{\partial R} + \frac{V_c}{R} \frac{\partial V_c}{\partial \theta} + \frac{U_c V_c}{R} - \frac{J_d}{Re} W_c \cos \theta = - \frac{1}{R} \frac{\partial X^*_c}{\partial \theta} \qquad (6.26)$$

$$U_c \frac{\partial W_c}{\partial R} + \frac{V_c}{R} \frac{\partial W_c}{\partial \theta} + \frac{J_d}{Re}\left[U_c \sin \theta + V_c \cos \theta \right] = - \frac{\partial X^*_c}{\partial Z} \qquad (6.27)$$

The secondary flow components in the core region are likely to be significantly smaller than that in the axial direction so that, taking U, V $\ll$ W in equations (6.25) through (6.27) we get

$$\frac{J_d}{Re} W_c \sin \theta = \frac{\partial X^*_c}{\partial R} \qquad (6.28)$$

$$\frac{J_d}{Re} W_c \cos \theta = \frac{1}{R} \frac{\partial X^*_c}{\partial \theta} \qquad (6.29)$$

$$U_c \frac{\partial W_c}{\partial R} + \frac{V_c}{R} \frac{\partial W_c}{\partial \theta} = - \frac{\partial X^*_c}{\partial Z} \qquad (6.30)$$

By eliminating the pressure-like terms from equation (6.28) and (6.29) it is possible to demonstrate that the core-region axial velocity field is functionally related to R sin θ which is the projection of the position vector of a typical point in the flow, viewed along the z-axis, onto a diameter joining the leading and trailing edges of the tube. The axial velocity in the core region is consequently made up from a series of ruled lines and has the form

$$W_c = f(R \sin \theta) \qquad (6.31)$$

where f is some function.
    In the core cross stream velocity components may be related to a stream function, Ψ^*_c by

$$RU_c = \frac{\partial \Psi^*_c}{\partial \theta} \qquad V_c = \frac{\partial \Psi^*_c}{\partial R} \tag{6.32}$$

Substitution of equations (6.31) and (6.32) into equation (6.30) and subsequent examination demonstrates that the stream function must have the form

$$\Psi^*_c = -\frac{R \cos \theta}{f'} \frac{\partial X^*_c}{\partial Z} \tag{6.33}$$

where

$$f' = \frac{df(R\sin\theta)}{d(R\sin\theta)} \tag{6.34}$$

Let us now examine equations (6.21) through (6.23) in the region very near to the tube wall where viscosity effects will be important. Suppose the thickness of this near-wall boundary layer, non-dimensionalised with respect to the radius of the tube is δ where $\delta \ll 1$. The non-dimensional axial velocity is of the order unity (written $W = O(1)$). Suppose the tangential velocity component in the boundary layer is of order $\varepsilon(V = O(\varepsilon))$, as yet undetermined, and take $\frac{\partial}{\partial Z} = O(1)$, $\frac{\partial}{\partial \theta} = O(1)$, $\frac{\partial}{\partial R} = O(\frac{1}{\delta})$. These assumed orders of magnitude may be combined with the continuity equation to show that the radial velocity component, U, must be of order $\varepsilon\delta(U = O(\varepsilon\delta))$.

Examination of equation (6.22) in the light of the orders of magnitude indicated above permits the following simplification to be effected in the boundary layer

$$U \frac{\partial V}{\partial R} + \frac{V}{R} \frac{\partial V}{\partial \theta} - \frac{UV}{R} - \frac{J_d}{Re} W \cos \theta = -\frac{1}{R} \frac{\partial X^*}{\partial \theta} + \frac{2}{Re} \frac{\partial^2 V}{\partial R^2} \tag{6.35}$$

$$O(\varepsilon^2) \quad O(\varepsilon^2) \quad O(\varepsilon^2\delta) \quad O(\frac{J_d}{Re}) \qquad\qquad\qquad O(\frac{\varepsilon}{Re\delta^2})$$

where the orders of magnitude are as indicated below some of the salient terms. To ensure that all the terms in equation (6.35) are of comparable order we see that

$$\varepsilon = O\left[\sqrt{\frac{J_d}{Re}}\right] \qquad \delta = O\left[\sqrt[4]{J_d \, Re}\right] \tag{6.36}$$

A similar examination of equation (6.23), noting that it has been assumed that $Re \gg J_d$, demonstrates that the axial pressure-type term may be ignored to give

$$U \frac{\partial W}{\partial R} + V \frac{\partial W}{\partial \theta} = \frac{2}{Re} \frac{\partial^2 W}{\partial R^2} \tag{6.37}$$

in the boundary layer region.

The core and boundary layer equations outlined above are valid provided the rotational speed is relatively low. The same procedure may be adopted for high rotational speeds which implies $J_d \gg Re$ and, in which case, inertial effects are negligible in the core and boundary layer regions. Equation (6.31) is consequently still valid in the core region. By omitting the inertia terms from equation (6.27) it may be shown that the stream function in the core must have the form

$$\Psi_c = \frac{Re}{J_d} \frac{\partial X^*c}{\partial Z} R \cos \theta \qquad (6.38)$$

The implication that inertial effects may also be ignored in the boundary layer at high rotational speeds reduces equation (6.35) to

$$-\frac{J_d}{Re} W \cos \theta = -\frac{1}{R} \frac{\partial X^*}{\partial \theta} + \frac{2}{Re} \frac{\partial^2 V}{\partial R^2} \qquad (6.39)$$

At high rotational speeds we may expect that the tangential velocity in the boundary layer is of similar order to that in the axial direction. This implies from continuity considerations that $U = O(\delta)$ and also, from equation (6.39) that $\delta = O(\frac{1}{J_d})$. An order of magnitude assessment of the axial pressure gradient-type term in equation (6.23) shows this term to be small in relation to the truncated viscosity terms and the Coriolis term so that axial momentum conservation in the boundary is satisfied by

$$J_d V \cos \theta = 2\frac{\partial^2 W}{\partial R^2} \qquad (6.40)$$

Benton and Boyer (1966) used the simplified core and boundary layer equation resulting from the assumptions of relatively high rotational speeds and solved them subject to the zero slip wall conditions. Their work was mainly involved with demonstrating the method of simplification for arbitrary cross sectional shapes rotating in the orthogonal mode and a detailed investiagion of the circular tube was not given.

Mori and Nakayama (1968) used the core and boundary layer region approach outlined above to study heat transfer and flow with the orthogonally rotating tube and effectively considered the secondary flow field to be strong in relation to the axial flow. The detailed analysis was similar to their earlier work with the parallel mode of rotation reported in Chapter 3.

With resistance coefficients defined in accordance with those used in equation (6.19) these authors derived the following result for the estimation of pressure loss

$$\frac{C_{fR}}{C_{fo}} = \frac{0.0992\left[\frac{J_d Re}{\Gamma}\right]^{-\frac{1}{4}}}{1 - 3.354\left[\frac{J_d Re}{\Gamma}\right]^{-\frac{1}{4}}} \qquad (6.41)$$

where

$$\Gamma = \left[1 + 0.3125\left[\frac{J_d}{Re}\right]^2\right]^{\frac{1}{2}} - 0.559\left[\frac{J_d}{Re}\right] \qquad (6.42)$$

The quantitative implications of equation (6.41) are demonstrated in figure 6.6 where the variation of flow resistance with rotational Reynolds number is shown for typifying values of through flow Reynolds numbers. Severe increases in resistance are evident and the impediment is increases as Reynolds number increases.

Ito and Nanbu (1971) took the core-region and boundary layer equations outlined earlier in this section and solved them using an integral method similar in principle to that used by Mori and Nakayama (1968). These authors treated the two cases resulting from the assumption that either $Re \gg J_d$ or $J_d \gg Re$. As a result of their analysis assuming that $Re \gg J_d$ they derived the following implicit result for flow resistance

$$\frac{C_{fR}}{C_{fo}} = \frac{0.08658(J_d Re)^{\frac{1}{4}}}{\left[1 - \frac{1.969}{\left[J_d Re\frac{C_{fR}}{C_{fo}}\right]^{1/5}}\right]^{5/4}} \qquad (6.43)$$

The implications of this equation are shown in figure 6.6 for the same values of through flow Reynolds numbers used to illustrate the result of the analysis of Mori and Nakayama (1968). The trends are very similar although precise numerical agreement is not evident.

For the case $J_d \gg Re$, Ito and Nanbu (1971) obtained the expression

$$\frac{C_{fR}}{C_{fo}} = \frac{0.0672\sqrt{J_d}}{\left[1 - \frac{2.11}{\sqrt{J_d}}\right]} \qquad (6.44)$$

and the implied variation of friction factor is also shown in figure 6.6. Although the trend is similar to the prediction of Mori and Nakayama (1968) the quantitative discrepancy is severe with equation (6.44) predicting flow resistance values approximately half of those of equation (6.41). It is interesting to note that equation (6.44) does not include the through flow Reynolds number on the right hand side of the equation and that, at the higher values of rotational Reynolds numbers, equation (6.41) also appears to become less dependent on through flow Reynolds number.

Skiadaressis and Spalding (1977) studied the problem of pipe flow with orthogonal-mode rotation using the numerical method they developed for parallel-mode rotation (see Skiadaressis and Spalding (1976)). These authors considered heated flow with either laminar or turbulent

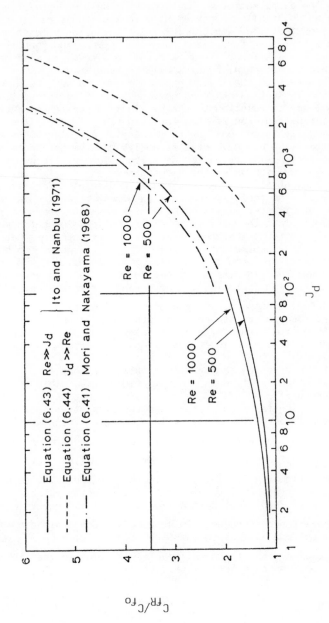

FIG. 6.6 THEORETICAL INFLUENCE OF ROTATION ON LAMINAR FLOW RESISTANCE. MORI AND NAKAYAMA (1968) AND ITO AND NANBU (1971).

flow and the method could, in principle, also solve the developing region. Although these authors do not appear to include the Coriolis-type terms in the axial momentum equation the resulting prediction were found to be physically in agreement with the trends resulting from the analyses described above. The friction factor predictions were compared with those of Mori and Nakayama (1968) and found to be typically about 8 - 14% lower. A more detailed discussion will be given later when experimental results will be described.

So far in this section the investigations discussed have been entirely theoretical. Let us now consider the theoretical predictions from these analyses in relation to the available experimental data.

The earliest reported experimental investigation of the influence of orthogonal-mode rotation on flow resistance in circular tubes was made by Trefethen (1957a) and, until recently, was the only data available. The theoretical investigation described above by Ito and Nanbu (1971) was augmented with an extensive programme of experiments with laminar and turbulent flow. Pressure losses and velocity profiles were measured with the radially inward flow of water in circular tubes having a range of diameter values. These measurements were made in tubes sufficiently long to justify the claim that developed flow was prevailing. It was found that the pressure loss measurements were in good agreement with accepted data at zero rotational speed. In this section the results for laminar flow will be discussed.

Figure 6.7 shows the experimentally determined effect of rotation on the Blasius friction factor, C_{fR}, for a range of rotational and through flow Reynolds numbers. In the usual laminar range there is a significant increase in resistance with rotation. Actual data points have been omitted from figure 6.7 in the interest of clarity but data scatter was generally low on individual lines of constant rotational Reynolds number.

The ratio of the friction factor with rotation to that for a stationary tube with laminar flow is shown in figure 6.8. This is an alternative method of presenting the same data as that shown in the previous figure. The friction factor is actually plotted against the product $J_d Re$ suggested by the analysis leading to equation (6.43) and lines of constant J_d values are shown. For each line of constant J_d value similar trends were evident and these are typified by the $J_d = 30$ line shown in figure 6.8. Initially increasing the Reynolds number value produces an attendent increase in friction factor ratio. At some level of the $J_d Re$ product a reduction in the ratio occurs with subsequent recovery and increase. This reduction was claimed to occur at the transition from laminar to turbulent flow. For each J_d-value tested it was claimed that the data immediately prior to transition tended to fall onto an asymptotic line given by

$$\frac{C_{fR}}{C_{fo}} = 0.0883 \, (J_d Re)^{1/4} \left[1 + 11.2 \, (J_d Re)^{-0.325} \right] \tag{6.45}$$

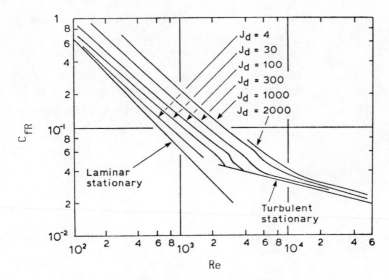

FIG. 6.7   EXPERIMENTALLY DETERMINED FRICTION FACTOR DATA
           WITH ROTATION.   ITO AND NANBU (1971).

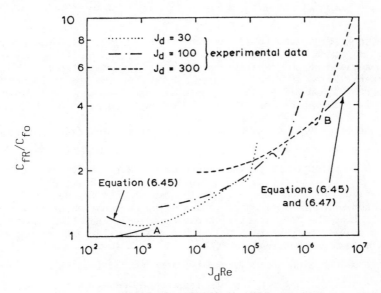

FIG. 6.8   EFFECT OF ORTHOGONAL MODE ROTATION ON LAMINAR
           FLOW RESISTANCE.   ITO AND NANBU (1971).
           (N.B. Equations (6.45) and (6.47) are almost
                 coincident in region AB and are omitted
                 for clarity).

valid in the range of variables $2.2 \times 10^2 \leqslant J_d Re \leqslant 10^7$ and $\dfrac{J_d}{Re} \leqslant 0.5$.
Below values of $J_d Re = 2.2 \times 10^2$ the influence of rotation was suff-
iciently small to be ignored in relation to the non-rotating case.
  As an alternative representation of the data these authors examined
the numerical coefficients in equation (6.43) in light of the experimental data and proposed the following modification, valid in the
range $J_d Re \geqslant 2 \times 10^3$ and $\dfrac{J_d}{Re} \leqslant 0.5$ for the asymptotic behaviour

$$\frac{C_{fR}}{C_{fo}} = \frac{0.0883 \, (J_d Re)^{1/4}}{\left(1 - \dfrac{1.969}{\left[J_d Re \dfrac{C_{fR}}{C_{fo}}\right]^{1/5}}\right)^{5/4}} \tag{6.46}$$

  This implicit equation could be expanded to an explicit form for
convenience of application to give

$$\frac{C_{fR}}{C_{fo}} = 0.0883(J_d Re)^{1/4} + 0.353 + 0.989(J_d Re)^{-1/4} +$$

$$+ 2.26(J_d Re)^{-1/2} + 4.40(J_d Re)^{-3/4} + O(\,(J_d Re)^{-1}) \tag{6.47}$$

  Equations (6.45) and (6.47) are compared with typical experimental
data in figure 6.8.
  A comparison of the experimental data for flow resistance of
Trefethen (1957a) and Ito and Nanbu (1971) with the predictions of
Mori and Nakayama (1968) and Skiadaressis and Spalding (1977) is given
in figure 6.9. Agreement with the numerical prediction of Skiadaressis and Spalding (1977) is very good.
  Ito and Nanbu (1971) also attempted to measure the axial developed
velocity profile in the plane of symmetry (i.e. $\theta = \dfrac{\pi}{2}$ and $\dfrac{3\pi}{2}$). Figure 6.10 compares their typical measurements with the theoretical
prediction of Skiadaressis and Spalding (1977) for Re = 4000, J_d =
1000. Although it appears curious to present data for Re = 4000 for
comparison with predictions of laminar flow examination of figure 6.7
suggests that the flow is still behaving in a laminar-like fashion.
It is probably the case that below this implied flow rate it was not
possible to make measurements.
  With the exception of the vicinity close to the leading edge ($\theta =$
$\dfrac{\pi}{2}$) the agreement between the predictions and experiment is good.

6.3 Isothermal Turbulent Flow in Circular-Sectioned Ducts

  Mori et al (1971) reported the results of a combined theoretical
and experimental investigation of flow resistance and heat transfer

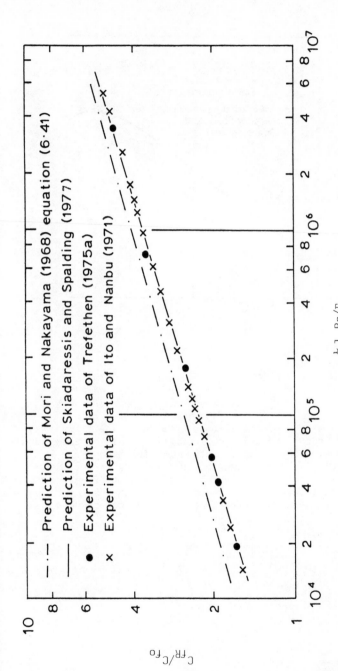

FIG. 6.9 COMPARISON OF THEORETICAL PREDICTIONS OF FLOW RESISTANCE WITH EXPERIMENTAL DATA FOR LAMINAR FLOW.

174

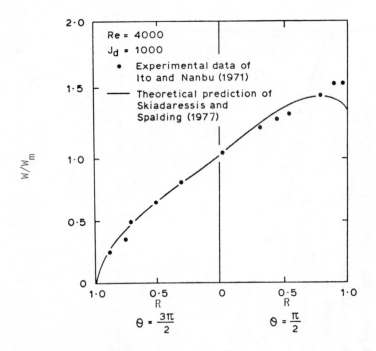

FIG. 6.10   COMPARISON OF THEORETICAL PREDICTIONS OF
AXIAL VELOCITY WITH EXPERIMENTAL DATA.

with turbulent flow and orthogonal-mode rotation. Their findings in
relation to flow resistance will be discussed in this section and the
heat transfer considered in later sections of this chapter.   In
essence Mori et al (1971) extended the earlier work of Mori and
Nakayama (1968) for laminar flow which itself was an extension of
their work on parallel-mode rotation and reported in Chapters 3 and 4
respectively.   The influence of rotation on flow resistance was determined to be

$$\frac{C_{fR}}{C_{fo}} = 0.962 \ X^{0.05}\left[1 + \frac{0.107}{X^{1.5}}\right] \qquad (6.48)$$

where

$$X = \frac{J_d^2}{4Re \ \Gamma^2}$$

$$\Gamma = \left[1 + 1.285 \ \frac{J_d^2}{Re^2}\right]^{\frac{1}{2}} - 1.135 \ \frac{J_d}{Re} \qquad (6.49)$$

The zero speed friction factor, C_{fo}, was determined from the usually accepted Blassius formula

$$C_{fo} = 0.316 \ Re^{-1/4} \qquad (6.50)$$

The implied flow impediment resulting from this analysis is shown in figure 6.11. The level of increase in friction factor is not as high as that experienced with laminar flow but, even so, is significant. The functional form of equation (6.48) is somewhat unwieldy and figure 6.12 does not directly illustrate the individual influences of the two Reynolds numbers used. These individual effects are more clearly shown in figure 6.12. As expected the rotation has a stronger effect, in relative terms, at lower values of through flow Reynolds number.

The numerical investigation reported by Skiadaressis and Spalding (1977), mentioned in the last section, also considered turbulent flow. Their prediction of flow resistance is shown plotted in figure 6.13. The abscissa used, namely J_d^2/Re, was selected as a result of the experimental findings of Ito and Nanbu (1971) for turbulent flow. The implications of this theoretical prediction appears to be that the curve shown is unique when plotted in the way shown. Also shown is the result of Mori et al (1971) for a range of rotational Reynolds number values. Equation (6.48) does not result in a unique curve when the resistance is plotted against J_d^2/Re and it is for this reason that the three separate sections, corresponding to J_d = 100, 300 and 1000 respectively, are presented in figure 6.13. It is interesting to note that the result of Skiadaressis and Spalding (1977) is in very good agreement with the experimental data of Ito and Nanbu (1971) and that the results of Mori et al (1971) appear to over predict the flow resistance.

The investigation of Ito and Nanbu (1971) also included a detailed experimental study of flow resistance with turbulent flow. Figure 6.14 typifies the experimental results noted. These authors found that they could correlate their turbulent flow resistance data against a parameter, J_d^2/Re, and this is shown in figure 6.14. At a fixed value of rotational Reynolds number it was found that the relative resistance to flow reduced as the through flow Reynolds number was increased. In this respect the zero speed friction factor was calculated using equation (6.50). As the Reynolds number, Re, was increased the flow resistance ratio achieved a minimum value after which further increases in Re tended to cause all data points to fall onto an asymptotic line as shown. This asymptote was deemed to correspond to a fully established turbulent flow regime by the authors with transition from laminar flow occurring at the location where serious departure from the asymtote occurred.

For the case of fully developed turbulent flow Ito and Nanbu (1971) proposed the following empirical correlation for flow resistance

$$\frac{C_{fR}}{C_{fo}} = 0.942 + 0.058 \left[\frac{J_d^2}{Re}\right]^{0.282} \qquad (6.51)$$

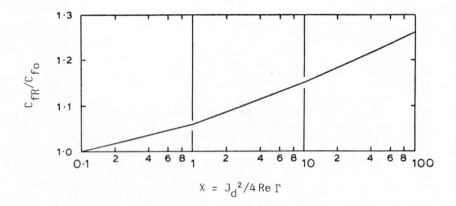

FIG. 6.11   INFLUENCE OF ROTATION ON TURBULENT FLOW
              RESISTANCE. MORI AND NAKAYAMA (1971).

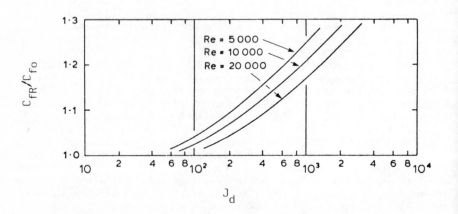

FIG. 6.12   ALTERNATIVE REPRESENTATION OF FLOW RESISTANCE
              PREDICTIONS FOR TURBULENT FLOW. MORI AND
              NAKAYAMA (1971).

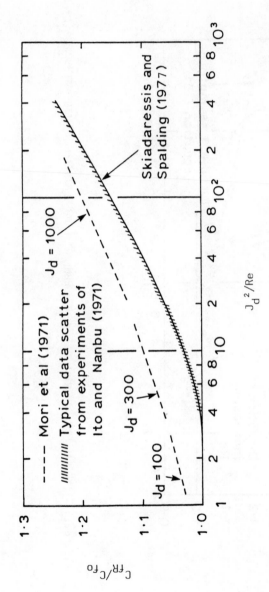

FIG. 6.13 COMPARISON OF THEORETICAL PREDICTIONS OF FLOW RESISTANCE WITH EXPERIMENTAL DATA FOR TURBULENT FLOW.

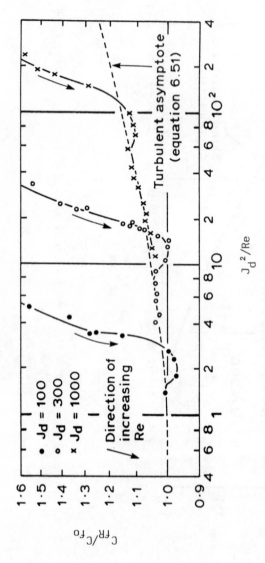

FIG. 6.14   TYPICAL EXPERIMENTAL DATA FOR TURBULENT FLOW RESISTANCE.   ITO AND NANBU (1971).

which is valid in the range $1 \leqslant J_d^2/\mathrm{Re} \leqslant 5 \times 10^5$.

Alternatively these authors proposed that for $J_d^2/\mathrm{Re} \geqslant 15$ flow resistance could be calculated using

$$\frac{C_{fR}}{C_{fo}} = 0.924 \left[\frac{J_d^2}{\mathrm{Re}}\right]^{1/20} \tag{6.52}$$

The fully developed turbulent data of Ito and Nanbu (1971 and Trefethen (1957b) are compared with equations (6.51) and (6.52) in figure 6.15. Visual observation suggests that the theoretical prediction of Skiadaressis and Spalding (1977) is in close agreement with equation (6.51).

Ito and Nanbu (1971) also attempted to measure the axial velocity profile in the plane of symmetry (i.e. $\theta = \frac{\pi}{2}$ and $\frac{3\pi}{2}$) and a typical result is compared with the theoretical predictions resulting from the numerical procedure of Skiadaressis and Spalding (1977) in figure 6.16. The comparisons are encouragingly good.

This section is concluded with a brief discussion on the transition from laminar to turbulent flow. Because their experiments were conducted with very disturbed flow conditions at the entry plane of their test section the occurrence of transition, as identified by curves similar to those shown in figure 6.14 corresponds to a lower critical through flow Reynolds number. Examination of their experimental data enabled Ito and Nanbu (1971) to propose the following equation for the transitional Reynolds number Re_c

$$\mathrm{Re}_c = 1070 \, J_d^{0.23} \tag{6.53}$$

Figure 6.17 illustrates the implication of equation (6.53) pictorially. There is a progressive delay in transition to turbulence as the rotational speed increases.

6.4 Laminar Heated Flow in Circular-Sectioned Ducts

6.4.1 Theoretical Studies of Developed Flow

Mori and Nakayama (1968) also included an analysis of laminar heat transfer in their investigation of orthogonal-mode rotation with uniformly heated circular-sectioned tubes. Their analysis again considers the influence of the Coriolis-induced secondary flow and the following results were developed for the heat transfer enhancement.

When the thermal boundary layer thickness is less than that of the hydrodynamic boundary layer (i.e. $\Delta_T \leqslant \Delta$), the fluid Prandtl number is greater than unity and the rotational Reynolds number, J_d, satisfies the constraint

$$J_d \leqslant \frac{\mathrm{Re}\,\mathrm{Pr}}{1.118}\left[1 - \frac{1}{\mathrm{Pr}^2}\right] \tag{6.54}$$

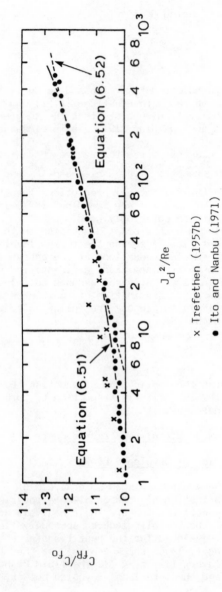

FIG. 6.15   COMPARISON OF DEVELOPED TURBULENT FLOW RESISTANCE DATA WITH EMPIRICAL
CORRELATIONS.   ITO AND NANBU (1971).

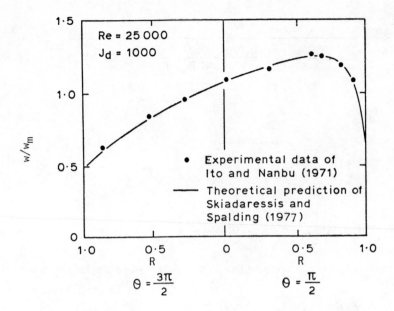

FIG. 6.16   COMPARISON OF THEORETICAL PREDICTIONS OF
AXIAL VELOCITY WITH EXPERIMENTAL DATA.

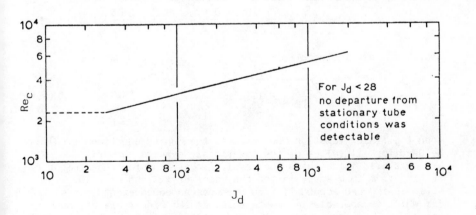

FIG. 6.17   EFFECT OF THE ROTATION ON THE TRANSITIONAL
REYNOLDS NUMBER. ITO AND NANBU (1971).

then

$$\frac{Nu_{\infty,b}}{Nu_o} = \frac{0.816\ X^{1/4}}{K\left[1 - 5.708\left[1 - \frac{0.176}{K} - \frac{0.07}{K\ Pr}\ (\frac{1}{K} + 4.430)\right]X^{1/4}\right]} \qquad (6.55)$$

where

$$X = J_d Re/\Gamma \qquad (6.56)$$

$$K = \frac{\Delta_T}{\Delta} = \frac{2}{11}\left[1 + \sqrt{\left(1 + \frac{77}{4}\ \frac{1}{Pr^2}\right)}\ \right] \qquad (6.57)$$

and

$$\Gamma = \left[1 + 0.315\left[\frac{J_d}{Re}\right]^2\right]^{1/2} - 0.559\left[\frac{J_d}{Re}\right] \qquad (6.58)$$

Alternatively when the rotational Reynolds number satisfies the constraint

$$J_d > \frac{Re\ Pr}{1.118}\left[1 - \frac{1}{Pr^2}\right] \qquad (6.59)$$

Mori and Nakayama (1968) show that heat transfer enhancement is given by

$$\frac{Nu_{\infty,b}}{Nu_o} = \frac{0.1816\ X^{1/4}}{K\left[1 - 3.354\left[K + \frac{0.5}{K} - \frac{0.1}{K^2} - \frac{0.8}{KPr}(K - \frac{0.25}{K} + \frac{0.05}{K^2})\right]X^{-1/4}\right]} \qquad (6.60)$$

where, in this instance

$$K = \frac{\Delta_T}{\Delta} = \frac{1}{5}\left[2 + \sqrt{\left(\frac{10}{\Gamma Pr^2} - 1\right)}\ \right] \qquad (6.61)$$

When the fluid Prandtl number is less than unity equation (6.60) may be used for all values of the rotational Reynolds number.

Figure 6.18 illustrates the level of heat transfer enhancement resulting from this analysis for the cases Pr = 2 and Pr = 10. The Coriolis-induced secondary flow produces an enhancement in heat transfer which is appreciable. Note that, at the higher Prandtl number value shown in figure 6.18, the discontinuity implied at the transition from the use of equations (6.55) and (6.60) is more marked.

These authors repeated the analysis but with an assumed constant wall temperature as the thermal boundary condition. It was found that the heat transfer coefficient obtained with rotation was relatively

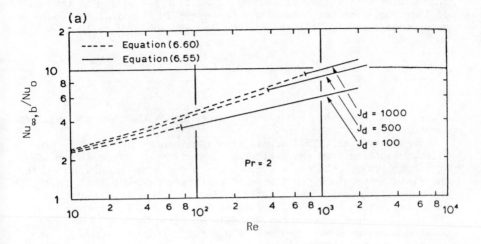

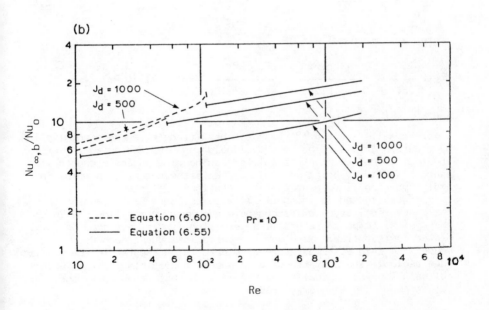

FIG. 6.18   EFFECT OF ROTATION ON LAMINAR HEAT TRANSFER.
MORI AND NAKAYAMA (1968).

insensitive to the thermal boundary condition prevailing at the wall
of the tube. The heat transfer enhancement in this instance could be
determined from the following simplified version of equations (6.55)
and (6.60)

$$\frac{Nu_{\infty,b}}{Nu_o} = 0.222 \frac{X}{K}^{1/4} \qquad (6.62)$$

where in this case $Nu_o = 3.66$; corresponding to the constant wall temperature boundary condition.

Vidyanidhi et al (1977) substituted the theoretical velocity profiles, resulting from the isothermal investigation of developed flow
of Barua (1955), into the energy equation and subsequently solved for
the fluid temperature profile (with uniformly heated walls). Although
the work of Vidyanidhi et al (1977) also includes the effect of viscous dissipation only non-dissipative results will be discussed here.
The heat transfer enhancement was defined using Nusselt numbers calculated using an unweighted mean fluid temperature difference. The
analysis produced the following result

$$\frac{Nu_{\infty,u}}{Nu_o} = \frac{1}{1 - 4\, Ja\, \Gamma} \qquad (6.63)$$

where, in this case

$$\Gamma = \frac{1}{24^2 . 20} + \frac{11\, Re_p^{\,2}}{96^2 . 8 . 2240} + \frac{Re_p^{\,2}\, Pr}{96^2 . 48^2 . 700}\left[3985 + 11310\, Pr\right] \qquad (6.64)$$

and $Nu_o = 6$.

For $Pr = 1$, figure 6.19 shows the heat transfer enhancement implied
by equation (6.64). The method of solution used, as noted in earlier
studies, is not all that useful for engineering purposes due to its
range of validity being so limited. Nevertheless it is helpful in
aiding physical interpretation of rotational effects.

Skiadaressis and Spalding (1977) also included heat transfer in
their numerical study of orthogonal-mode rotation and figure 6.20 shows
the typical temperature profiles resulting from a developed uniformly
heated laminar flow. Note how the Coriolis-induced secondary flow
causes the largest temperature gradients to be located in the trailing
edge region. The heat transfer enhancement predicted by this numerical solution was typically 15% higher than that proposed by the analysis of Mori and Nakayama (1968).

6.4.2 Experimental Studies of Laminar Heat Transfer

The earliest recorded attempt to experimentally determine the effect
of orthogonal-mode rotation on heat transfer was made by Mori, Fukada
and Nakayama (1971). This investigation also extended the theoretical
predictions of Mori and Nakayama (1968) for laminar flow to conditions

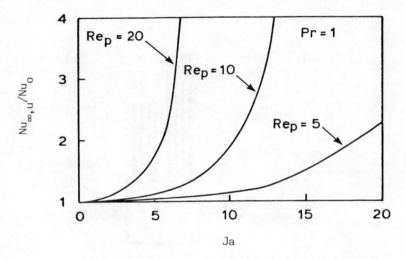

FIG. 6.19   THEORETICAL INFLUENCE OF CORIOLIS-INDUCED
SECONDARY FLOW ON HEAT TRANSFER.   VIDYANIDHI
ET AL (1977).

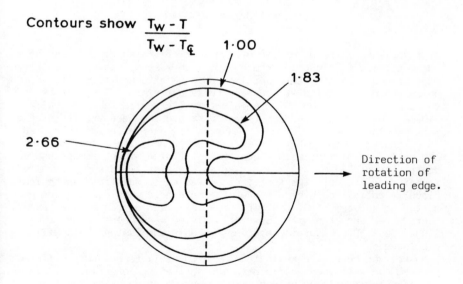

FIG. 6.20   TYPICAL TEMPERATURE DISTRIBUTION FOR LAMINAR
FLOW.   SKIADARESSIS AND SPALDING (1977).
(Re = 1300; J_d = 78).

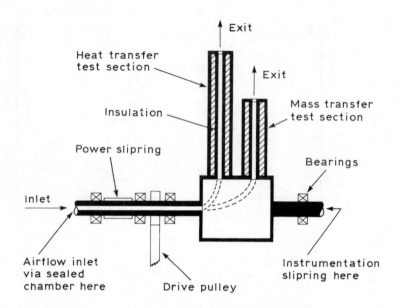

FIG. 6.21   SCHEMATIC ARRANGEMENT OF APPARATUS USED BY
MORI, FUKADA AND NAKAYAMA (1971).

of turbulence. The schematic arrangement of the apparatus used by
these authors is shown in figure 6.21. In essence either a heat transfer or mass transfer test section could be attached to a rotor system
which could be rotated at speeds upto 1000 rev/min. The heat transfer
measurements where made with an electrically heated tube nominally
510 mm long with a bore diameter of 9 mm and the eccentricity of the
mid-span location was 380 mm. The mass transfer experiments deemed
to be analogous to heat transfer were performed with a test section
nominally 320 mm long with a bore diameter of 6 mm and a mid-span eccentricity of 285 mm. Air was used as the test fluid and for the mass
transfer experiments the test section wall was made from naphthalene.
Circumferential measurements over the tube bore with naphthalene permitted an assessment of the girthwise variation of convection. Experiments were conducted with flow rates giving a Reynolds number range
1000 < Re < 10,800 but in this section only the laminar data will be
discussed.

Nusselt numbers were avaluated using the bulk temperature difference
between the wall and the fluid as the motivating potential for heat
transfer and developed values were evaluated at an axial location 63%
along the active length of the test sections.

The results of the heat transfer and analogous mass transfer tests
are compared with equations (6.60) and (6.62) respectively in figure
6.22. The heat transfer experimental data was higher than that pre-

187

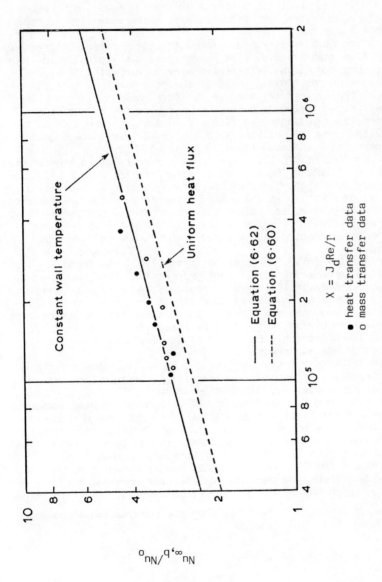

FIG. 6.22 COMPARISON OF THEORETICAL DEVELOPED LAMINAR HEAT AND MASS TRANSFER WITH EXPERIMENTAL DATA FROM MORI, FUKADA AND NAKAYAMA (1971) WITH AIR AS THE TEST FLUID.

dicted by Mori and Nakayama (1968) for the case of uniformly heated tubes and actually fell closer to their constant wall temperature predictions. Since the uniform heat flux prediction of Skiadaressis and Spalding (1977) was about 15% higher than that of Mori and Nakayama (1968) it is evident that the prediction from the numerical procedure appears most accurate.

The variation of Nusselt number around the circumference of the tube calculated from mass transfer experiments is shown typically in figure 6.23. Note how the girthwise heat transfer is highest in the local region located on the trailing edge of the tube and that this is consistent with the typical temperature contours shown in figure 6.20. The theoretical variation of Nusselt number over the circumference for the conditions indicated in figure 6.23 and determined using the predictive technique of Mori and Nakayama (1968) was in good qualitative agreement with the mass transfer analogue.

Metzger and Stan (1977) also reported the results of an experimental investigation of laminar heat transfer with parallel-mode rotation. These authors used a transient method whereby a disc with a series of radial holes machined in it was heated prior to rotation at prescribed speeds and with prescribed coolant flow rates fed to the radial holes. The range of parameters studied is shown in Table 6.1 where for comparison the corresponding data for Mori, and Fukada and Nakayama (1971) is also shown.

The experimental data of Metzger and Stan (1977) was found to give heat transfer enhancement significantly lower than that suggested by the available theoretical analyses. This may be seen on reference to figure 6.24 where the enhancement measured is compared with the prediction of Mori and Nakayama (1968). In this respect the enhancement ratio involves an average value for the entire length of the tubes studied and is not strictly the ratio of developed values. Nevertheless the trends are important.

When parallel-mode rotation was treated in earlier chapters the buoyant interaction between the centripetal field an a temperature dependent fluid density was found to be an important consideration. It is important to note that buoyancy effects have not been considered in any of the theoretical models postulated to date for orthogonal-mode rotation. Only the beneficial effect of Coriolis acceleration on heat transfer have been included in all the analyses. Let us now attempt to assess the likely effect of buoyancy in this case.

Consider again the flow and coordinate system shown in figure 6.1. If the fluid reference temperature for the assessment of buoyancy is taken as T_0, which may be the uniform temperature of the fluid entering the heated section of the tube, then the combination of equation (2.50) with equation (6.2) yields

$$\frac{Dv}{Dt} + 2(\underline{\omega} \wedge \underline{v}) = -\frac{1}{\rho_0} \nabla p' + \upsilon \nabla^2 \underline{v} +$$

$$+ \frac{1}{2} \Omega^2 \beta \ (T - T_0) \ \nabla \left[r^2 \sin^2 \theta + (H + z)^2 \right] \qquad (6.65)$$

as the vectorial form of the laminar Navier-Stokes equation.

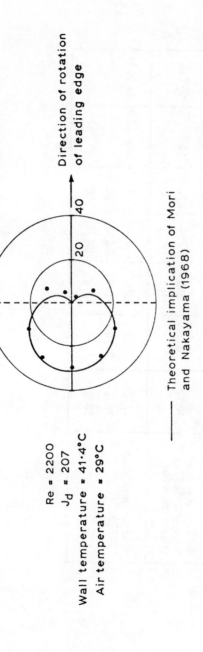

FIG. 6.23 TYPICAL VARIATION OF LOCAL NUSSELT NUMBER WITH LAMINAR FLOW FROM NAPHTHALENE-AIR MASS TRANSFER MEASUREMENTS OF MORI, FUKADA AND NAKAYAMA (1971).

Source	L (mm)	d (mm)	H (mm)	L/d	H/d	Re	J_d
Mori, Fukada and Nakayama (1971) (Heat Transfer)	510	9	380	56.67	42.22	1000+	0-500
Mori, Fukada and Nakayama (1971) (Mass Transfer)	320	6	285	53.33	47.50	1000+	0-240
Metzger and Stan (1977)	38.1	6.35	31.8	6	5	700-3100	0-700
	38.1	3.18	31.8	12	10		
	38.1	1.59	31.8	24	20		

TABLE 6.1   RANGE OF EXPERIMENTAL VARIABLES FOR LAMINAR FLOW WITH ORTHOGONAL-MODE ROTATION.

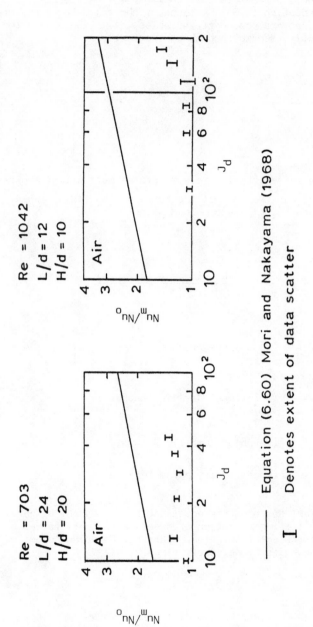

FIG. 6.24   COMPARISON OF EXPERIMENTAL DATA OF METZGER AND STAN (1977) WITH
PREDICTIONS OF MORI AND NAKAYAMA (1968) FOR LAMINAR FLOW.

The non-dimensional parameters influencing the flow field may be brought out from equation (6.65) by means of the following transformation of independent and dependent variables (all symbols have the meanings consistently adopted throughout the text).

$$R = r/a \quad , \quad Z = z/a$$

$$V = \frac{v}{w_m} \quad , \quad P' = \frac{P'}{\rho_o w_m^2} \quad , \quad \eta = \frac{T-T_o}{Pr \, \Delta T_w} \tag{6.66}$$

$$\underline{\Psi} = \frac{\omega}{\Omega}$$

Note that ΔT_w is a representative measure of the motivating temperature difference between the wall of the tube and the fluid. When equations (6.65) and (6.66) are combined we get

$$\frac{Dv}{Dt} + Ro(\underline{\Psi} \wedge \underline{V}) = - \nabla P' + \frac{2}{Re} \nabla^2 \underline{V}$$

$$- \frac{1}{16} \frac{Ra_b}{Re} \eta \, \nabla \left[\frac{R^2 \sin^2 \theta + (2\varepsilon + Z^2)}{\varepsilon} \right] \tag{6.67}$$

where

$$Ro = \frac{\Omega d}{w_m} = \frac{J_d}{Re} \qquad \text{(Rossby number)}$$

$$Ra_b = \frac{H\Omega^2 \beta d^3 \, \Delta T_w \, Pr}{\upsilon^2} \qquad \text{(Rotational Rayleigh number)}$$

$$Re = \frac{w_m d}{\upsilon} \qquad \text{(Reynolds number)} \tag{6.68}$$

$$\varepsilon = H/d \qquad \text{(Eccentricity parameter)}$$

The Coriolis effect has been characterised by the Rossby number, Ro, above to suit some later description of experimental data to be reported on heated turbulent flow. The rotational Reynolds number, J_d, could easily be used also if necessary since

$$Ro = \frac{J_d}{Re} \tag{6.69}$$

Equation (6.67) shows that the grouping Ra/Re^2 is an important parameter in assessing the influence of buoyancy. When the flow is radially outwards the buoyancy situation is analogous to a vertical tube influenced by the earth's field and having a downward flow. The

The axial buoyancy component thus opposes the customary forced convection resulting in an expected impediment to heat transfer. This suggests that, with a radially outward axial flow in the tube, the rotational buoyancy will tend to oppose the beneficial effects of the Coriolis acceleration. The converse might be expected with radially inward flow.

Unfortunately there is insufficient information available to make a precise assessment of centripetal-buoyancy effects but one explanation for the relatively low heat transfer enhancement reported by Metzger and Stan (1977) could be the impediment resulting from opposed buoyancy. This feature will be discussed in more detail in the next section when the results of an investigation of turbulent heat transfer in general and buoyancy in particular will be treated.

6.5 Turbulent Heated Flow in Circular-Sectioned Ducts

6.5.1 Theoretical Studies of Developed Flow

The momentum integral technique has also been used by Mori, Fukada and Nakayama (1971) to estimate the enhancement in heat transfer due to the Coriolis acceleration with orthogonal-mode rotation. For fluids having a Prandtl number of about unity or less these authors propose that

$$\frac{Nu_{\infty,b}}{Nu_o} = \frac{1.026 \ Pr^{2/3} \ X^{1/20}}{[Pr^{2/3} - 0.074]} \left(1 + \frac{0.093}{X^{1/5}} \right) \qquad (6.70)$$

where

$$X = \frac{J_d^{\,2}}{4 \ Re \ \Gamma^2}$$

$$\Gamma = \left[1 + 1.285 \ \frac{J_d^{\,2}}{Re^2} \right]^{1/2} - 1.135 \ \frac{J_d}{Re} \qquad (6.71)$$

and the zero speed Nusselt number is calculated using

$$Nu_o = 0.034 \ Re^{3/4} \qquad (6.72)$$

When the fluid has a Prandtl number greater than unity the proposal is that

$$\frac{Nu_{\infty,b}}{Nu_o} = 1.087 \ X^{1/30} \left[1 + \frac{0.059}{X^{1/6}} \right] \qquad (6.73)$$

where in this case

$$\chi = \frac{0.177 \ J_d^{2 \cdot 5}}{Re^{1 \cdot 5} \ \Gamma^{2 \cdot 5}}$$

$$\Gamma = \left[1 + 1.623 \ \frac{J_d^2}{Re^2} \right]^{1/2} - 1.275 \ \frac{J_d}{Re}$$

(6.74)

and

$$Nu_o = 0.023 \ Re^{0 \cdot 8} \ Pr^{0 \cdot 4}$$

(6.75)

It is not easy to readily discern the influence of rotation from the mathematical structure of either equation (6.70) for (6.73) but reference to figures 6.25 and 6.26 aids physical appreciation. In figure 6.25 the heat transfer enhancement implied by this analysis is shown for a range of Re and J_d values with air as the test fluid. The Coriolis effect produces significant improvement in heat transfer but, as expected, the enhancement is not as relatively high as with laminar flow. For a specified value of rotational Reynolds number the enhancement decreases as the through flow Reynolds number increases. Similar physical trends emerge from equation (6.73) for fluids having a Prandtl number greater than unity as shown in figure 6.26.

The numerical method of Skiadaressis and Spalding (1977) suggested much lower heat transfer enhancement than the prediction of Mori, Fukada and Nakayama (1971) and this is depicted in figure 6.27. In the analysis of Skiadaressis and Spalding (1977) an influence due to Reynolds number over and above that resulting from the analysis of Mori, Fukada and Nakayama (1971) is also evident.

The two attempts to analyse turbulent flow with orthogonal-mode rotation described above represent, as far as this author is aware, the only available sources of information. Attention will now be focussed on experimental data which is currently available.

6.5.2 Experimental Studies of Turbulent Heat Transfer

A number of experimental studies of heat transfer with orthogonal-mode rotation have been reported and these are summarised in Table 6.2 Mori, Fukada and Nakayama (1971) also conducted a limited number of experiments with turbulent flow and figure 6.28 compares the results of their heat and mass transfer results with their own prediction based on equation (6.70). Although insufficient information was given to identify the individual influence of Re and J_d effects the agreement appeared to be reasonable. Figure 6.29 shows the circumferential variation of Nusselt number typically measured from their mass transfer tests. The angular variation was not found so severe as that found with laminar flow.

Lokai and Limanski (1975) also conducted a series of experiments with heated radially outward flow. Figure 6.30 summarises the results noted. Despite wide data scatter these authors noted improved heat

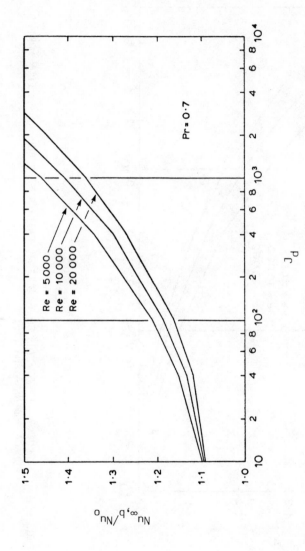

FIG. 6.25 INFLUENCE OF ROTATION ON TURBULENT HEAT TRANSFER FOR AIR FROM MORI, FUKADA AND NAKAYAMA (1971).

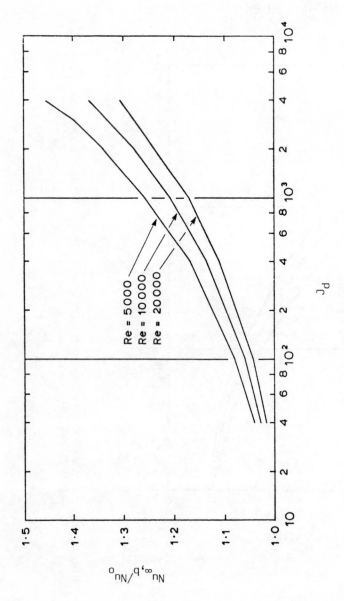

FIG. 6.26   INFLUENCE OF ROTATION ON TURBULENT HEAT TRANSFER FOR $Pr > 1$ FROM MORI, FUKADA AND NAKAYAMA (1971).

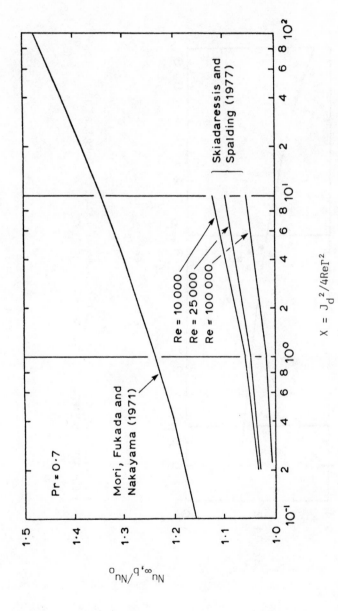

FIG. 6.27   COMPARISON OF THEORETICAL STUDIES OF DEVELOPED TURBULENT HEAT TRANSFER WITH AIR.

198

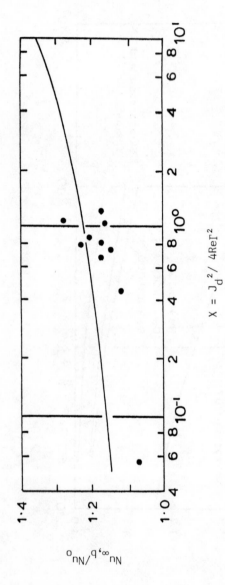

$X = J_d{}^2 / 4Re\Gamma^2$

FIG. 6.28 COMPARISON OF THEORETICAL AND EXPERIMENTAL DATA FOR TURBULENT HEAT TRANSFER WITH AIR FROM MORI, FUKADA AND NAKAYAMA (1971).

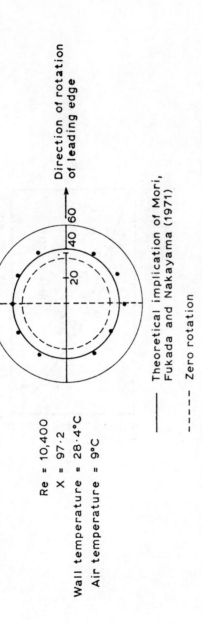

Re = 10,400

X = 97·2

Wall temperature = 28·4°C

Air temperature = 9°C

Direction of rotation
of leading edge

20 40 60

——— Theoretical implication of Mori,
Fukada and Nakayama (1971)

----- Zero rotation

FIG. 6.29   TYPICAL VARIATION OF LOCAL NUSSELT NUMBER WITH TURBULENT FLOW FROM NAPH-
THALENE-AIR MASS TRANSFER MEASUREMENTS OF MORI, FUKADA AND NAKAYAMA (1971).

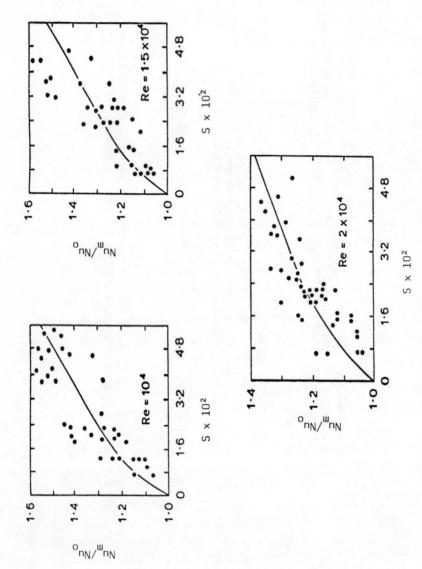

FIG. 6.30   EXPERIMENTALLY DETERMINED HEAT TRANSFER DATA FOR TURBULENT FLOW FROM LOKAI AND LIMANSKI (1975).

transfer and proposed the following correlating equation

$$\frac{Nu_m}{Nu_o} = \left[1 + 136.97 \, \frac{Ro^{0.7}}{Re^{0.38}} \right] = \left[1 + 136.97 \, \frac{J_d^{0.7}}{Re^{1.08}} \right] \qquad (6.76)$$

where Nu_o was determined from an appropriate zero speed turbulent correlation of heat transfer. Lokai and Limanski (1975) used the Rossby number, S, to represent the Coriolis effect in their original work and equation (6.76) shows the equivalent form in terms of the rotational Reynolds number. Figure 6.31 compares the implications of using the prediction of Mori, Fukada and Nakayama (1971) and the correlation of Lokai and Limanski (1975) for two representative Reynolds numbers and the agreement between the two proposals is not good.

Zysina-Molozhen, Dergach and Kogan (1977) undertook a programme of experiments with heated radially outward flow and figure 6.32 is a synopsis of their findings. These authors suggest that the flow tends to exhibit a more laminar-like behaviour when rotation is present and that at the higher end of their Reynolds number range the rotation has no serious effect (say Re > 2 x 10^4). In the region where rotation was deemed to be important the enhancement in heat transfer was correlated by an equation which, in terms of present nomenclature has the structure

$$\frac{Nu_m}{Nu_o} = 20.476 \left\{ 1 - \frac{94}{\left[J_d \, \frac{H}{d} + 3600 \right]^{0.7}} \right\} Re^{-0.3} \qquad (6.77)$$

The experiments conducted by these authors appear to have involved an eccentricity $\frac{H}{d}$ = 32.5 and based on this assumption the enhancement for Re = 10 000 implied by equation (6.77) is compared with the proposals of Mori, Fukada and Nakayama (1971) and Lokai and Limanski (1975) in figure 6.31. The use of equation (6.77) gives results which are significantly lower than those implied by the use of equation (6.70) or (6.76).

An examination of the heat transfer data led Zysina-Mologhan, Dergach and Kogan (1977) to propose the following correlation for the transition from laminar to turbulent flow when heated flow is considered.

$$Re_c = 2300 \left\{ 1 + 0.0026 \left[J_d \, \frac{H}{d} \right]^{0.84} \right\} \qquad (6.78)$$

This equation is compared with the isothermal proposal of Ito and Nanbu (1971) in figure 6.33 for the case $\frac{H}{d}$ = 32.5. Serious discrepancies between the two results are clearly in evidence suggesting a marked effect of combined heating and rotation on the transitional Reynolds number.

202

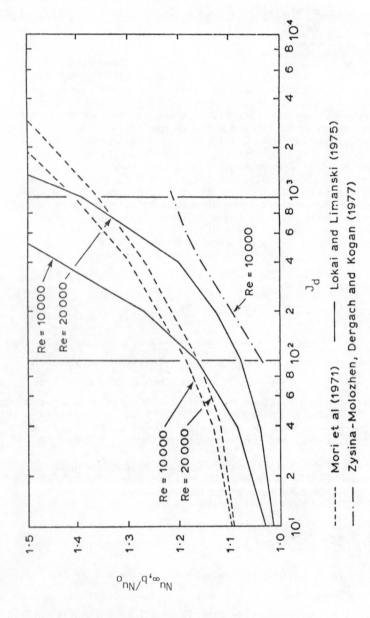

FIG. 6.31 COMPARISON OF VARIOUS PROPOSALS FOR ESTIMATING TURBULENT HEAT TRANSFER
ENHANCEMENT.

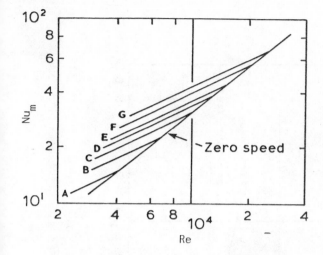

d = 6mm
L = 150mm
H ≈ 195mm

	Ω (rev/min)
A	60
B	250
C	500
D	750
E	1000
F	1500
G	2150

FIG. 6.32   EXPERIMENTALLY DETERMINED TURBULENT HEAT
TRANSFER FROM ZYSINA-MOLOZHEN, DERGACH
AND KOGAN (1977).

As in the case of laminar flow treated earlier in this chapter the
investigations of heated turbulent flow and orthogonal-mode rotation
have all assumed that Coriolis acceleration is the sole manifestation
of rotation. The theoretical and experimental results have not, it
would appear, been in complete accord. It is evident that the data
scatter experienced by the experimental studies currently available
has been partially the result of buoyancy effects as briefly discussed
in section 6.4.2. This led the present author to undertake a systematic experimental programme of heat transfer experiments designed
specifically to isolate the influence of Coriolis and centripetal
buoyancy effects and this work will now be discussed. Full details
are available from Morris and Ayhan (1979). The range of experimental
variables is shown in Table 6.2 together with those of other studies
of turbulent flow. A simplified line diagram of the apparatus used
is shown in figure 6.34, which, in essence, consisted of a rotor arm
to which could be attached various heated test sections in the appropriate attitude. Two test sections were used for the present tests.
Each comprised a stainless steel tube nominally 100.00 mm or 150 mm
in length and with a corresponding bore diamter of either 4.85 mm or
10.00 mm. These test sections will be designated A and B respectively.
The test sections were located on the rotor arm so that the eccentricity of the mid-span position was 306 mm for test section A and 328
mm for test section B.
Heating of the active lengths of the test sections was achieved
electrically using Nichrome resistance wire spirally wound over the
outer periphery of the tubes. Current was fed to the heater via a
power slip ring mounted on the main shaft and controlled with a

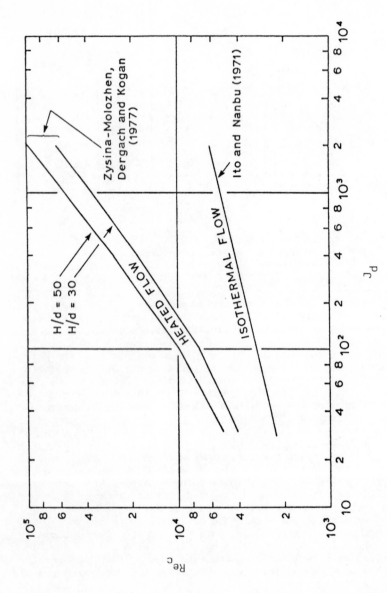

FIG. 6.33   INFLUENCE OF ROTATION OF THE TRANSITIONAL REYNOLDS NUMBER WITH ISOTHERMAL
AND HEATED FLOW.

Source	L (mm)	d (mm)	H (mm)	L/d	H/d	Re	J_d	Ra
Mori, Fukada & Nakayama (1971) (Heat Transfer)	510	9	380	56.67	42.22	1000	0 – 500	?
Mori, Fukada & Nakayama (1971) (Mass Transfer)	320	6	285	53.33	47.50	1000	0 – 240	?
Lokai and Limanski (1975)	130 approx	5 approx	?	26	?	$10^4 – 2 \times 10^4$	0 – 600	?
Zysina-Molozhen, Dergach and Kogan (1977)	150	6	195	25.00	32.50	$4 \times 10^3 – 3 \times 10^4$	0 – 526	?
Morris & Ayhan (1979) (Radially Outward Flow)	A 100 B 150	4.85 10	306 328	20.62 15.00	63.09 32.80	$5 \times 10^3 – 15 \times 10^3$ $5 \times 10^3 – 15 \times 10^3$	0 – 300 0 – 1300	$2 \times 10^4 – 6 \times 10^5$ $5 \times 10^5 – 7 \times 10^6$
Morris & Ayhan (1981) (Radially Inward Flow)	C 100 D 150	4.85 10	306 328	20.62 15.00	63.09 32.80	$5 \times 10^3 – 15 \times 10^3$ $5 \times 10^3 – 15 \times 10^3$	0 – 300 0 – 1300	$5 \times 10^4 – 7 \times 10^5$ $7 \times 10^5 – 9 \times 10^6$

TABLE 6.2   RANGE OF EXPERIMENTAL VARIABLES FOR TURBULENT FLOW WITH ORTHOGONAL-MODE ROTATION.

206

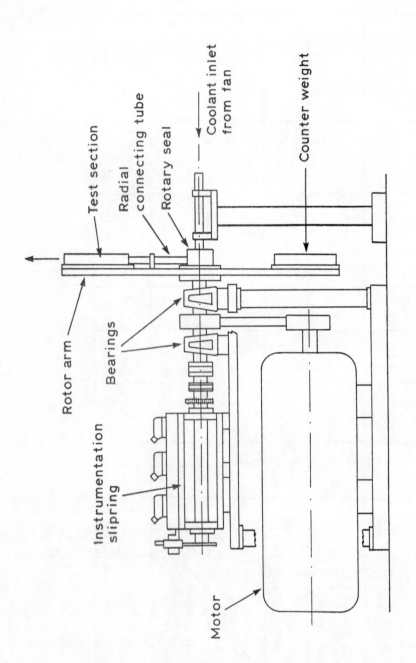

Test section

Radial
connecting tube

Rotary seal

Coolant inlet
from fan

Counter weight

Rotor arm

Bearings

Instrumentation
slipring

Motor

FIG. 6.34   SCHEMATIC ARRANGEMENT OF APPARATUS USED BY MORRIS AND AYHAN (1979).

variable transformer. The outer surface of the test section was covered with a layer of asbestos cord to inhibit external heat loss and the entire section covered with a protective shell of stainless steel. Thermocouples were mounted along the effective length of the test sections to permit an accurate measurement of wall temperature distribution. Thermocouples also permitted the measurement of the air temperature at entry to and exit from the test section. At the exit end a series of baffles in the tube enable the bulk temperature of the fluid to be more readily assessed. All rotor-mounted thermocouple signals were monitored with a data logger with signals taken from the rig via a silver-silver graphite instrumentation slip ring assembly.

The main rotor assembly comprising main shafting, rotor arm and heated test section was mounted between self-aligning bearings and driven, at speeds up to 2000 rev/min by means of a controlled electric motor. Rotational speeds were measured magnetically using a timer counter.

The air used for the test fluid was blown through the test section in the radially outward direction with a centrifugal blower. The flow circuit comprised a mass flow meter, inlet rotary seal assembly and radial connecting duct.

Experiments were conducted with both test sections at rotational speeds of 0, 1000 and 2000 rev/min. This gave a maximum centripetal acceleration of about 1400g at the mid-span location of the test sections. Flow rates giving nominal Reynolds number values of 5000, 10 000 and 15 000 were treated and for each speed-Reynolds number combination a range of heater dissipation rates were examined.

For a specified run the power dissipation by the heater is not totally transferred to the coolant owing to inevitable external heat loss. These losses may be broadly attributed to conduction at each end of the test section together with an external convective-type loss. The external convective loss is dependent on the rotational speed and this was estimated from a series of calibration tests undertaken prior to the main experimental programme.

From steady state measurements of the flow rate, power dissipation, rotational speed, temperature levels, etc., it was possible to determine the non-dimensional groups expected to govern the system.

Figure 6.35 shows, for the case of zero rotational speed and test section A, typical variations of local Nusselt number, Nu_z, measured along the test section. In this respect the local Nusselt number is defined in terms of the local heat flux and the locally prevailing difference in wall and bulk coolant temperature, with properties evaluated at the bulk temperature of the fluid. The following features are worthy of note for these validation tests.

For each of the nominal Reynolds numbers used the local heat transfer variations were similar in that there was an initial region of high Nusselt number with a tendency to decrease towards an asymptotic fully developed value. In accordance with typical ducted flow forced convection experiments there is a region towards the end of the test section where exit-type effects produce an upswing in the local Nusselt number. Also shown in figure 6.35 are the fully developed values of Nusselt number suggested by the well-known Dittus-Boelter (1930) correlation for turbulent flow in circular-sectioned tubes. Agreement

208

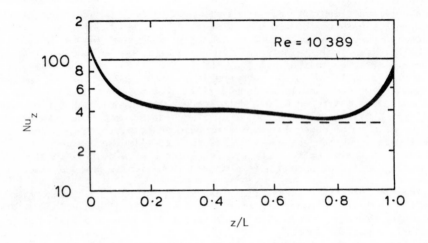

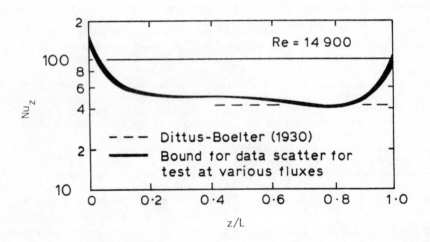

FIG. 6.35   TYPICAL AXIAL VARIATION OF LOCAL NUSSELT
NUMBER AT ZERO ROTATIONAL SPEED FOR TEST
SECTION A FROM MORRIS AND AYHAN (1979).

with the Dittus-Boelter proposal was generally found to be good in the most developed regions of the test section.

The experimental data shown in figure 6.35 were produced from tests having a variety of heat flux levels with each Reynolds number value. Although each heat flux level produces different wall to fluid temperature relationships for a specified Reynolds number there should be a tendency for all data to collapse onto a single curve for the zero rotational speed case when expressed in non-dimensional form. This was found to be true and the dark bands shown in figure 6.35 bound the variations resulting from all heat flux tests. This tendency for the non-rotational data to collapse into tight bands is important and will be discussed further when the non-zero rotational speed data are considered. Local heat transfer measurements made with test section B also followed a similar trend.

Figure 6.36 illustrates local variations in Nusselt number resulting from rotation of test section A at 1000 rev/min. The individual curves shown for each of the three nominal Reynolds number values represent different heat flux conditions. These individual curves did not demonstrate a tendency to collapse into a tight band as had been the case at zero speed. From the earlier discussion concerning the physical manifestations of rotation it was expected that centripetal buoyancy would impair convection. Examination of the results for each Reynolds number value shown in figure 6.36 demonstrated that the reduction in local heat transfer was systematically related to increases in the heat flux conditions, that is to the level of operating temperature. Because the experiments were controlled to ensure that individual variations in Reynolds number over any nominal Reynolds number series of tests was minimal (typically about 1 - 2 per cent) the systematic reduction in heat transfer observed with increases in heat flux cannot be attributed to viscosity increases with temperature.

Over any of the three series of tests shown in figure 6.36 only the rotational Rayleigh number varied significantly in response to different heat flux settings thus substantiating the expected influence of centripetal buoyancy. Identical observations to those described above were also noted at a rotational speed of 2000 rev/min and also with the experiments repeated using test section B.

The influence of rotation on mean heat transfer is typically demonstrated in figure 6.37. Here the ratio of mean Nusselt number, Nu_m, to that assuming a stationary tube and the same Reynolds number is plotted against the Reynolds number itself for data obtained with test section A and a rotational speed of 1000 rev/min. At relatively low heating rates, where the wall-bulk fluid temperature differences were relatively low, the heat transfer was improved in relation to the stationary tube situation. In this region of enhanced heat transfer the proposals of Nakayama et al (1971), Lokai and Limanski (1975). Skiadaressis and Spalding (1977) and Zysina et al (1977) are also shown. Generally the data from the present tests at low heating rates tended to show closer agreement with the theoretical prediction of Skiadaressis and Spalding (1977).

However as the heat flux level and hence the Rayleigh number was raised there was a systematic reduction in the relative mean Nusselt number in accordance with the reduction in local Nusselt number demon-

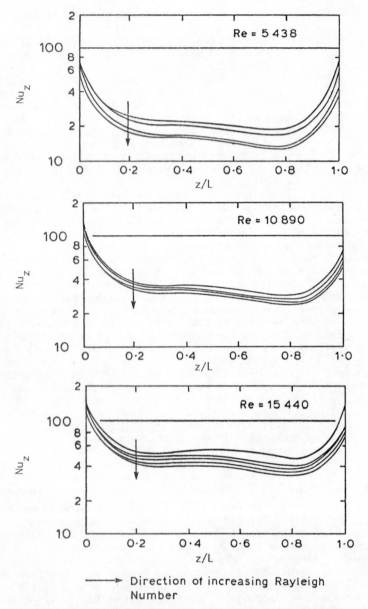

Direction of increasing Rayleigh
Number

FIG. 6.36   EFFECT OF ROTATION ON LOCAL TURBULENT HEAT
TRANSFER FROM MORRIS AND AYHAN (1979) WITH
RADIALLY OUTWARD FLOW.   (TEST SECTION A,
ROTATIONAL SPEED = 1000 REV/MIN).

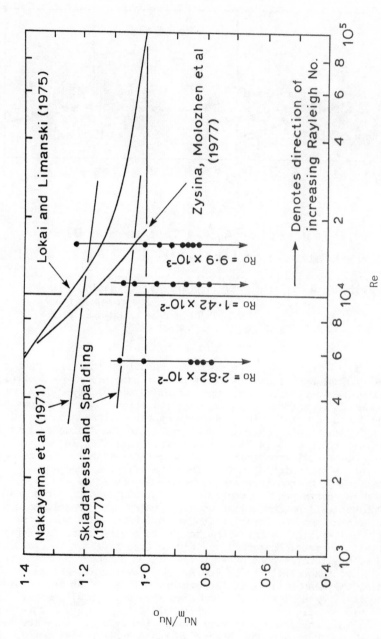

FIG. 6.37   INFLUENCE OF ROTATION ON TURBULENT RELATIVE MEAN NUSSELT NUMBER WITH RADIALLY OUTWARD FLOW FROM MORRIS AND AYHAN (1979). (TEST SECTION A, ROTATIONAL SPEED = 1000 REV/MIN).

212

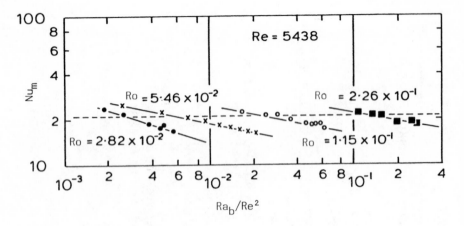

(•,× Test section A, ◦,■ Test section B)
----- Dittus-Boelter (1930)

FIG. 6.38 INFLUENCE OF CORIOLIS ACCELERATION AND
CENTRIFUGAL BUOYANCY ON MEAN TURBULENT
HEAT TRANSFER FOR RADIALLY OUTWARD FLOW
FROM MORRIS AND AYHAN (1979).

strated in figure 6.36. It should be re-emphasised that, for each of
the three nominal Reynolds numbers shown in figure 6.37, the control
of the experiments was such that the actual values of Reynolds and
Rossby numbers did not depart from their respective nominal values by
more than 1 - 2 per cent. Thus Rayleigh number is the only variable
parameter for each set. An important feature in the design context
to note from figure 6.37 is the serious over-prediction in heat transfer which may result from the use of theoretically and experimentally
derived predictions which do not include the effect of buoyancy.
    Examination of the non-dimensional groups suggested by equation
(6.67) and the corresponding energy equations suggests that the mean
Nusselt number will depend on the usual Reynolds and Prandtl numbers
together with either the Rossby or rotational Reynolds number, the
quotient of a Rayleigh number with the square of the Reynolds number
together with geometric parameters. Figure 6.38 typifies the results
obtained with all rotational speeds, flow rates and heating rates when
plotted in accordance with these ideas. The Rayleigh number was evaluated using the mean wall-fluid temperature difference as the motivating potential for heat transfer and also the mean Nusselt number.
    This figure highlights clearly the twofold manner in which rotation
of the tube influences the heat transfer. Firstly at a specified Reynolds number and Rossby number, which is tantamount to fixing the
peripheral speed of the coolant channel, the mean level of heat trans-

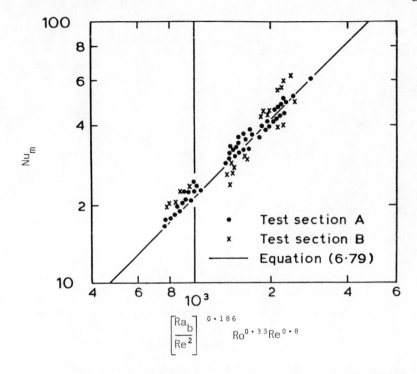

$$\left[\frac{Ra_b}{Re^2}\right]^{0.186} Ro^{0.33}Re^{0.8}$$

FIG. 6.39   COMPARISON OF PROPOSED CORRELATION OF MORRIS
AND AYHAN (1979) WITH EXPERIMENTAL DATA FOR
RADIALLY OUTWARD FLOW.

fer is systematically reduced as the quotient Ra/Re^2 increases.   The
rotationally induced free convection produces a significant impediment
in the heat transfer for this condition of radial outflow since the
buoyant motion is in opposition to the usual forced convection.
At a fixed value of the Reynolds and Rayleigh number the heat transfer increases with increases in the Rossby number, which is a measure
of the relative influence of the Coriolis force.
Although the geometric features of test sections A and B imply differences in length/diameter ratio and eccentricity parameter, it was
nevertheless found possible to correlate mean Nusselt numbers obtained
at all rotational speeds with a simple exponent-type relationship involving the Reynolds number, Rossby number and Rayleigh number along
the lines suggested by figure 6.38.   Thus with a mean scatter band of
±15 per cent all rotating data was correlated by

$$Nu_m = 0.022\ Re^{0.8}\left[\frac{Ra_b}{Re^2}\right]^{-0.186}Ro^{0.33}\tag{6.79}$$

and figure 6.39 illustrates the location of all data points with respect to this equation.   Because of the complexity of the rotational

interaction, it was not found possible with the data currently available, to propose a simple correlating equation which reduces to the
stationary tube case at zero rotational speed.

For air over the range of length/diameter ratios used in these experiments the Kreith (1965) proposal for mean Nusselt number, Nu_0, in
a stationary tube is well approximated by the result

$$Nu_0 = 0.022 \, Re^{0.08} \qquad (6.80)$$

which conveniently enables equation (6.79) to be re-expressed as

$$\frac{Nu_m}{Nu_0} = \left[\frac{Ra_b}{Re^2}\right]^{-0.186} Ro^{0.33} = M \qquad (6.81)$$

Although equation (6.81) is restricted to the range of non-dimensional variables covered in the test programme reported (see Table
6.2 for details) it is nevertheless useful for assessing whether, at
a particular operating condition, one is likely to encounter a region
of enhanced or impaired heat transfer relative to the stationary tube
situation. Figure 6.40 shows all data points resulting from the present study plotted in the manner suggested by equation (6.81). Despite
the data scatter present we see that for $M > 1$, the overall effect of
rotation is to enhance the customary pipe flow heat transfer whereas
the converse is true for $M < 1$.

It is not strictly possible to make direct comparison with the results obtained with radially outward flow and the analogous situation
in a vertical tube influenced by the earth's gravity since it is not
possible to isolate buoyancy resulting from a centripetal field from
other Coriolis induced effects. For example there is evidence that
an expected increase in heat transfer due to gravitational buoyancy
with an upward heated turbulent flow can be nullified and reversed
due to relaminarisation of the flow near the wall. The fact that in
the present case of radially outward flow discussed above that the
heat transfer diminishes suggests that the flow is behaving in a more
laminar fashion. Useful background reading on gravitational buoyancy
superimposed onto forced convection in vertical tubes may be found in
the works of Skeele and Hanratty (1962), Hall and Jackson (1969),
Herbert and Sterns (1972), Byrne and Ejiogu (1972) and Easby (1976).

In order to check out the effect of centripetal buoyancy in more
detail Morris and Ayhan (1981) repeated their entire programme of experiments for the case of radially inward flow for now it was expected
that buoyancy ideally should improve heat transfer.

Figure 6.41 shows the variation of local Nusselt number measured
along test section C and typifies all the results obtained. Tests at
zero rotational speed had again demonstrated the same trends as shown
in figure 6.35. With radially inward flow it is seen on reference to
figure 6.41 that there is now a systematic increase in local heat
transfer as the heat flux levels and hence the buoyancy effect increases. Figure 6.42 shows the corresponding effect of rotation on
the enhancement in mean Nusselt number. At the three values of nomi-

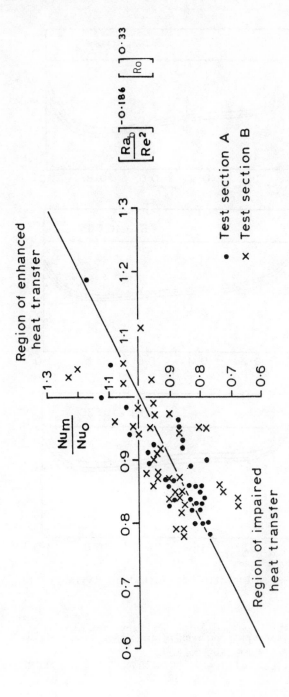

FIG. 6.40 PROPOSED BOUNDS FOR IMPAIRED AND ENHANCED HEAT TRANSFER WITH RADIALLY OUTWARD FLOW FROM MORRIS AND AYHAN (1979).

216

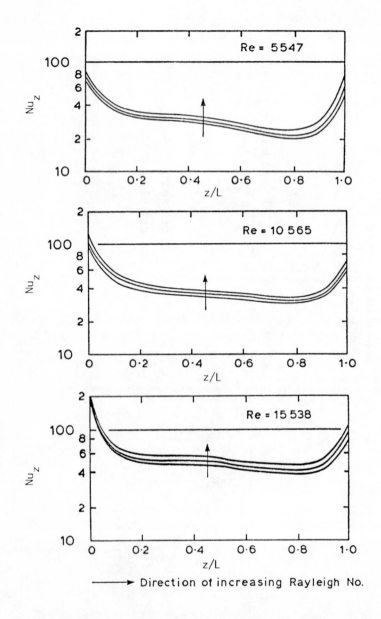

Direction of increasing Rayleigh No.

FIG. 6.41   EFFECT OF ROTATION ON LO1AL TURBULENT HEAT
TRANSFER FROM MORRIS AND AYHAN (1981) WITH
RADIALLY INWARD FLOW.
(TEST SECTION C, ROTATIONAL SPEED = 1000 RPM).

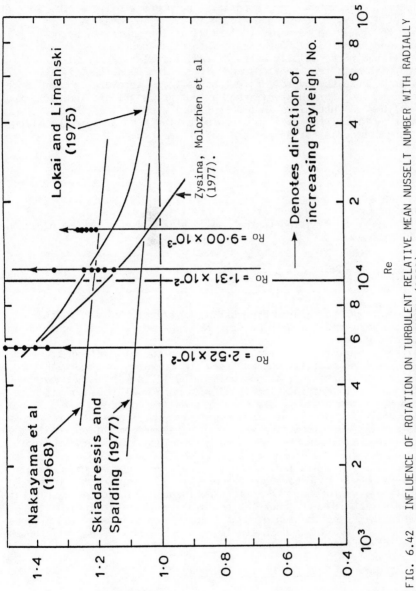

FIG. 6.42 INFLUENCE OF ROTATION ON TURBULENT RELATIVE MEAN NUSSELT NUMBER WITH RADIALLY
INWARD FLOW FROM MORRIS AND AYHAN (1981).
(TEST SECTION C, ROTATIONAL SPEED = 1000 REV/MIN).

nal Reynolds numbers studied the significant enhancement in heat transfer resulting from the overall influences of rotation is clearly in
evidence. Also shown for comparative purposes are the predictions of
Mori et al (1971), Lokai and Limanski (1975), Skiadaressis and
Spalding (1977) and Zysina et al (1977). The empirical proposal of
Lokai and Limanski (1975) appears marginally better even though it
was derived with radially outward flow. Similar trends were again
noted with both test sections and all running conditions tested. Figure 6.43 shows the overall effect of Coriolis acceleration and centripetal buoyancy for the typifying case of Re = 5521 plotted along the
same lines as that given in figure 6.38.

A study of all the data obtained permitted the following equation
to be constructed for correlation of the mean Nusselt number

$$Nu_m = 0.036 \ Re^{0 \cdot 8} \left[\frac{Ra_b}{Re^2} \right]^{0 \cdot 112} Ro^{-0 \cdot 083} \tag{6.82}$$

The enhancement in heat transfer resulting from equation (6.82) may
be expressed in relation to the non-rotating case by use of equation
(6.80) to give

$$\frac{Nu_m}{Nu_o} = 1.72 \left[\frac{Ra_b}{Re^2} \right]^{0 \cdot 112} Ro^{-0 \cdot 083} = N \tag{6.83}$$

All data was correlated by equation (6.82) with a mean scatter of
± 13 per cent and all data points are compared with this equation in
figure 6.44.

Although Coriolis acceleration was shown to improve heat transfer
in the case of radially outward flow the converse is true with the inward flow situation. This unexpected result is not fully understood
at present but certainly implies a much more complex interaction between the rotational effects then originally conceived.

The following points are made concerning the effect of rotation on
a notionally turbulent flow with orthogonal-mode rotation. All theoretical investigations to date have not included buoyancy interactions
between a temperature dependent fluid density and the centripetal acceleration. This has led to the doubtful feature that theoretical
predictions all tend to suggest improved heat transfer due to the
Coriolis acceleration. There is experimental evidence available which
raises serious doubts about the reliability of not including buoyancy
effects in the estimation of heat transfer under these circumstances
although the available data itself does not fully explain all the physical interactions present.

6.6 Recommendations for Design Purposes

This chapter has demonstrated that there are still a number of
serious doubts and uncertainties in the understanding of orthogonal-
mode rotation and its effect on convective processes. Nevertheless
the following comments offer advice and guidance when design-type

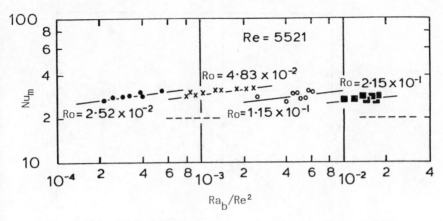

(•, × Test section C, ○, ■ Test section D)

----- Dittus-Boelter (1930)

FIG. 6.43   INFLUENCE OF CORIOLIS ACCELERATION AND
CENTRIFUGAL BUOYANCY ON MEAN TURBULENT
HEAT TRANSFER FOR RADIALLY INWARD FLOW
FROM MORRIS AND AYHAN (1981).

estimates of flow resistance and heat transfer are needed.

Flow Resistance with Unheated Flow

Orthogonal-mode rotation causes secondary flow to be created even
with fully developed flow. With laminar flow the developed predictions of Mori and Nakayama (1968) (see equation (6.41)) and Ito and
Nanbu (1971) (see equation (6.43)) are in reasonable agreement and
are useful for estimation purposes. The modified version of equation
(6.43) resulting from the experiments of Ito and Nanbu (1971) (see
equation (6.47)) is useful for estimating flow resistance with laminar
flow in the range $J_d Re \geqslant 2 \times 10$ and $J_d/Re \leqslant 0.5$.

When the flow is turbulent the predictions of Skiadaressis and
Spalding (1977) appear to give good agreement with the experimental
data of Ito and Nanbu (1971) but their theoretical result is not explicitly available in mathematical terms. Accordingly the empirically-
based equation (6.51) is particularly useful for calculating turbulent
flow resistance in the range $1 \leqslant J_d^2/Re \leqslant 5 \times 10^5$ or alternatively

equation (6.52) for the range $J_d^2/Re \geqslant 15$.

For unheated flow no distinction between radially inward or outward
flow need be made since the Coriolis acceleration only tends to re-
verse direction. Because of the uncertainty described earlier due to
buoyancy effects care should be exercised in estimating flow resistance with heated flows and no definitive suggestion can be made at

220

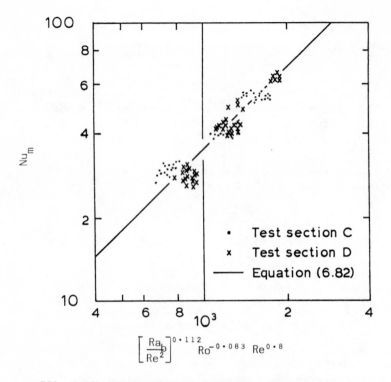

$$\left[\frac{Ra_b}{Re^2}\right]^{0.112} Ro^{-0.083} Re^{0.8}$$

FIG. 6.44   COMPARISON OF PROPOSED CORRELATION OF
MORRIS AND AYHAN (1981) WITH ALL
EXPERIMENTAL DATA FOR RADIALLY INWARD
FLOW.

present.

It is evident that orthogonal-mode rotation tends to suppress the transition from laminar to turbulent flow. The isothermal transitional Reynolds number may be estimated using equation (6.53). This equation was empirically determined with radially inward flow and it is not clear with available data whether it may be confidently used with radially outward flow. Zysina-Molozhan, Dergach and Kogan (1977) have proposed equation (6.78) for the determination of transition with heated radially outward flow. The results of this equation imply a much more stable behaviour in comparison to equation (6.53).

Heat Transfer with Laminar Flow

All theoretical studies of heat transfer with orthogonal-mode rotation have only included the effect of Coriolis acceleration and consequently should only be used in situations where buoyancy effects are known to be small. This implies low rates of heat transfer with relatively low motivating temperature differences between the wall of the tube and the fluid. For these cases the theoretical study of Mori and Nakayama (1968) gives a method of assessment which may be used to a first approximation via equations (6.55), (6.60) and (6.62). Although no data has been reported which explicitly investigates the buoyancy effect in laminar flow the M and N-functions derived by Morris and Ayhan (1979), (1981) for turbulent flow may be used as a guide for radially outward and inward flow respectively.

Heat Transfer with Turbulent Flow

Uncertainty also exists in the estimation of turbulent flow heat transfer particularly in relation to the effect of buoyancy. With currently available data it is recommended that the theoretical studies of Mori, Fukada and Nakayama (1971) and Skiadaressis and Spalding (1977) are used in conjunction with the empirical proposals of Lokai and Limanski (1975) and Zysina-Molozhan, Dergach and Kogan (1977). This implies the use of equations (6.70), (6.73), (6.76) and (6.77) respectively in the first instance. This approach should be followed by an assessment of buoyancy using the empirical equations (6.81) and (6.83) for radially outward and inward flow respectively. It is unfortunately not possible to be more precise, more detailed new investigations need to be undertaken to clearly establish the inter-relationships between Coriolis acceleration and centripetal buoyancy.

It is interesting to conclude this chapter with a discussion of the implication of the results reported on the performance of spanwise gas turbine rotor blade coolant channels. It has often been reported (see for example Koehler, Henneke et al (1977)) that blade temperatures measured in the real engine situation are higher than those predicted either from simulated cascade studies or theoretically estimated using external boundary layer models together with a solution of the conduction equation in the blade material. Although a number of contributory reasons may be possible for these discrepancies, it should be noted that incorrectly specified boundary conditions at the surfaces of cooling holes can significantly influence the resulting blade temperature. Recall the statement of Fox (1974) mentioned in Chapter 1 that a ± 10% variation on coolant hole heat transfer results in a ± 20°C change in mean blade temperature under typical operating conditions and that this implies a severe reduction in blade life if the heat transfer variation is negative.

Table 6.3 cites coolant hole geometries, rotational speed and cooling air conditions which typify operating conditions for an industrial gas turbine and a modern aircraft fan engine respectively. Also shown are the implications on the heat transfer relative to the zero rotational speed condition resulting from the calculation procedures discussed in this chapter. The serious over prediction in heat transfer resulting from the use of correlations which do not take into account buoyancy with the radially outward flow is clear.

Engine Type	Industrial gas turbine	Aircraft fan-type gas turbine
GEOMETRIC DETAILS:		
Coolant passage length (mm)	50.0	90.0
Coolant passage diameter (mm)	5.8	1.7
Midspan eccentricity (mm)	305.0	386.0
Rotational speed (rev/min)	10 000	10 000
COOLANT FLOW DETAILS:		
Mean pressure (bar)	6.0	10.0
Mean temperature (K)	603	702
Mean axial velocity (m/s)	114.0	308.0
Mean blade temperature (K)	333	1073
NON-DIMENSIONAL PARAMETERS		
Length/diameter ratio	8.6	52.9
Eccentricity parameter	52.6	227.1
Flow Reynolds number	7.47×10^4	7.61×10^4
Rotational Reynolds number	3.98×10^3	4.40×10^2
Rossby number	5.33×10^{-2}	5.78×10^{-3}
Rotational Rayleigh number	3.15×10^8	1.63×10^7
Rayleigh/Reynolds number^2	5.65×10^{-2}	2.81×10^{-3}
PERCENTAGE CHANGE IN RELATIVE MEAN NUSSELT NUMBER		
Mori, Fukada and Nakayama (1971) (Radially outward and inward flow)	+44.6	+22.0
Lokai and Limanski (1975) (Radially outward and inward flow)	+24.7%	5.2%
Zysina-Molozhan, Dergach and Kogan (1977) (Radially outward and inward flow)	0%	0%
Morris and Ayhan (1979) (Radially outward flow)	-35.0%	-46%
Morris and Ayhan (1981) (Radially inward flow)	+60%	+36%

TABLE 6.3 DESIGN IMPLICATIONS OF EXTRAPOLATING TRENDS
REPORTED TO REAL ENGINE OPERATING CONDITIONS.

BIBLIOGRAPHY

Arnold,J.J.(1970). Mechanical problems in large rotating electrical machines. Chart. Mech. Engr. Dec. 478.

Barua,S.N.(1955). Secondary flow in a rotating straight pipe. Proc. Roy. Soc. A, 227, 133.

Bayley,F.J., Owen,J.M. and Turner,A.B.(1972). Heat Transfer. Thomas Nelson Publishing Co., London.

Bennett,R.B.(1968). Water cooling of turbine generator rotor windings. Engl. Elec. Jour. 23, 18.

Benton,G.S., and Boyer,D.(1966). Flow through a rapidly rotating conduit of arbitrary cross-section. J.Fluid Mech. 26, part 1, 69.

Boussinesq. (1930). Theorie analytique de la chaleur. Gathiers-Villars, Paris. 2.

Creek,F.R.L.(1977). Progress in the design of large turbo-generators. Turbine Generator Engineering, A.E.I. Turbine Generators Ltd., Trafford Park, Manchester 17, 117.

Coriolis,G.G.(1829). Traite de la mecanique solides.

Davies,T.H. and Morris,W.D.(1966). Heat transfer characteristics of a closed loop rotating thermosyphon. Proc. Third Int. Heat Transfer Conf., A.I.Ch.E., Chicago, USA. 2, 172.

Dias,F.M.(1978). Heat transfer and resistance to flow in rotating square tubes. D.Phil. Thesis, University of Sussex, Falmer,England.

Dittas,F.W., and Boelter,L.M.K.(1930). Univ. Calif.Publs.Engng. 2, 443.

Fox,M.(1974). Some implications of heat transfer in turbine blade design: a framework for comparison of design strategies. Aero. Res. Council Report AR35250, HMT 339.

Goldstein,S.(1957). Modern Developments in Fluid Dynamics. Oxford University Press, London. Vol. 2.

Gosman,A.D., Pun,W.M., Runchal,A.K., Spalding.D.B.and Wolfshtein,M.W. (1969). Heat and Mass Transfer in Re-circulating Flows. Academic Press.

Harlow,F.H., and Nakayama,P.I.(1967). Turbulent transport equations. Physics of Fluids, 10 (11), 2323.

Hawley,R.(1969). Turbo-type generators of the future. Elec. Times. 12 June, 6.

Humphreys,J.F.(1966). Convection heat transfer in a revolving tube. Ph.D. Thesis, University of Liverpool, Liverpool.

Humphreys,J.F., Morris,W.D. and Barrow, H.(1967). Convection heat transfer in the entry region of a tube which revolves an axis parallel to itself. Int. Jour. Heat Mass. Trans. 10, 333.

Ito,H., and Nanbu,K.(1971). Flow in rotating straight pipes of circular cross section. ASME Trans., J. Basic Eng. 93 (3), 383.

Koehler,H., Hennecke,D.K., Pfoff,K., and Eggebrecht,R.(1977). Hot cascade test results of cooled turbine blades and their application to actual engine conditions. Proc. 50th Agard-PEP Symposium on High Temperature Problems in Gas Turbines. Ankara, Turkey.

Lamb,H.(1945). Hydrodynamics. 6th Edition. Dover Publications, New York.

Launder,B.E., and Spalding,D.B.(1974). The numerical computations of turbulent flows. Comp. Methods in Appl. Mech. and Eng. $\underline{3}$, 269.

Le Feuvre.R.(1968). Heat transfer in rotor cooling ducts. Thermo and Fluid Mech. Convention, I.Mech.E., Bristol. Proc. I.M.E. $\underline{182}$, (3), 232.

Lighthill,M.J.(1949). A technique for tendering approximate solutions to physical problems uniformly valid. Phil. Mag. $\underline{40}$, 1179.

Lokai,V.I., and Limanski,A.S.(1975). Influence of rotation on heat and mass transfer in radial cooling channels of turbine blades. Izvestiya VUZ, Aviatsionnaya Tekhika, $\underline{18}$, No. 3, 69.

Majumdar,A.K., Morris,W.D., Skiadaressis,D., and Spalding,D.B.(1977). Heat transfer in rotating ducts. Mech.Eng.Bull., Central Mech. Eng. Research Inst., Durgapur, $\underline{8}$ (4), 87.

Metzger,D.E., and Stan,R.L.(1977). Entry region heat transfer in rotating radial tubes. AIAA 15th Aerospace Sciences Meeting, Los Angeles. Paper No. 77-189.

Mori,Y., Fukada,T., and Nakayama,W.(1971). Convective heat transfer in a rotating radial circular pipe (2nd report). Int.J.Heat Mass Transfer, $\underline{14}$, 1807.

Mori,Y., Futagami,K., Tokuda,S., and Nakayama,W.(1966). Forced convective heat transfer in uniformly heated horizontal tubes. Int. J. Heat Mass Trans. $\underline{9}$, 453.

Mori,Y., and Nakayama,W.(1967). Forced convective heat transfer in a straight pipe rotating about a parallel axis (laminar region). Int. J.Heat Mass Trans. $\underline{10}$, 1179.

Mori,Y., and Nakayama,W.(1968). Convective heat transfer in rotating radial circular pipes (1st report, laminar region). Int.J.Heat Mass. Transfer. $\underline{11}$, 1027.

Morris,W.D.(1964). Heat transfer characteristics of a rotating thermo-syphon. Pd.D. Thesis, University of Wales, Swansea.

Morris,W.D.(1965a). The influence of rotation on flow in a tube rotating about a parallel axis with uniform angular velocity. Jour. Roy. Aero Soc. $\underline{69}$, 201.

Morris,W.D.(1965b). Laminar convection in a heated vertical tube rotating about a parallel axis. Jour. Fluid Mech. $\underline{21}$, Part 3, 453.

Morris,W.D.(1968). Terminal laminar convection in a uniformly heated rectangular duct. Thermo and Fluid Mech. Convention, I.Mech.E., Bristol. Paper No. 4.

Morris,W.D.(1969). An experimental investigation of laminar heat transfer in a uniformly heated tube rotating about a parallel axis. Min. Tech. ARC CP No. 1055.

Morris,W.D.(1970). Some observations on the heat transfer characteristics of a rotating mixed convection thermosyphon. Min. Tech. ARC CP No. 1115.

Morris,W.D.(1981). A pressure transmission system for flow resistance measurements in a rotating tube. J. Phys: Sci. Instrum. $\underline{14}$, 208.

Morris,W.D., and Ayhan,T.(1979). Observations on the influence of rotation on heat transfer in the coolant channels of gas turbine rotor blades. Proc. Inst. Mech. Eng. $\underline{193}$, No.21, 303.

Morris,W.D., and Ayhan,T.(1981). Heat transfer in a rotating tube with radially inward flow. University of Hull, Department of Engineering Design and Manufacture. Report No. EDM/4/81.

Morris,W.D., and Dias,F.M.(1981). Laminar heat transfer in square-sectioned ducts which rotate in the parallel-mode. Power Ind. Res. $\underline{1}$.

Morris,W.D., and Woods,J.L.(1978). Heat transfer in the entrance region of tubes that rotate about a parallel axis. Jour. Mech. Eng. Sci. $\underline{20}$ (6), 319.

Morton,B.R.(1959). Laminar convection in uniformly heated horizontal pipes at low Rayleigh numbers. Quart. Jour. Mech. Appl. Math. $\underline{12}$, 410.

Nakayama,W.(1968). Forced convective heat transfer in a straight pipe rotating about a parallel axis (turbulent region). Int. Jour. Heat Mass Trans. $\underline{11}$, 1185.

Nakayama,W., and Fuzioka,K.(1978). Flow and heat transfer in the water-cooled rotor winding of a turbine generator. I.E.E.E. Trans. Power App. and Systems. PAS $\underline{97}$ (1), 225.

Nusselt,W.(1910). Z.Ver. deut. Ingr., 54: 1154-1158.

Pohl,H.(1973). Development of the two-pole turbo alternator with full liquid cooling. Brown Boveri Review 2/3, 85.

Sakamoto,M., and Fukui,S.(1971). Convective heat transfer of a rotating tube revolving about an axis parallel to itself. Elec. and Nuclear Eng. Lab., Tokyo Shibaura Elec. Co. Ltd., Kawasaki, Japan.

Schlichting,H.(1968). Boundary Layer Theory. 6th Edition. McGraw Hill Book Co. Inc., New York.

Seider,E.N., and Tate,G.E.(1936). Heat transfer and pressure drop of liquids in tubes. Ind. Eng. Chem. $\underline{28}$, 1429.

Siegworth,D.P., Wikesell,R.D., Reodal,T.C., and Hanratty,T.J.(1969). Effect of secondary flow on the primary temperature field and primary flow in a heated horizontal tube. Int. Jour. Heat Mass Transfer. $\underline{12}$, 1535.

Skiadaressis,D., and Spalding,D.B.(1976). Laminar heat transfer in a pipe rotating about a parallel axis. Imperial Coll. Science and Tech. Mech. Eng. Report No. HTS/76/23.

Skiadaressis,D., and Spalding,D.B.(1977). Heat transfer in a pipe rotating around a perpendicular axis. ASME Paper No. 77-WA/HT-39.

Trefethen,L.(1957a). Laminar flow in rotating radial ducts. Report R55GL350/55GL350-A. General Electric Company.

Trefethen,L.(1957b). Fluid flow in radial rotating tubes. Actes, 1X. Congres International de Mecanique Appliquee, Universite de Bruxelles, 2, 341.

Vidyanidhi,V., Suryanarayana,V.V.S., and Chenchu,R.(1977). An analysis of steady fully developed heat transfer in a rotating straight pipe. Trans. ASME Jour. Heat Transfer, Feb. 148.

Woods,J.L.(1975). Heat transfer and flow resistance in a rotating duct system. D.Phil. Thesis, University of Sussex, Falmer, England.

Woods,J.L., and Morris,W.D.(1974). An investigation of laminar flow in the rotor windings of directly-cooled electrical machines. J. Mech. Eng. Sci. 16, 408.

Woods,J.L., and Morris,W.D.(1981). A study of heat transfer in a rotating cylindrical tube. Trans. ASME, J. Heat Trans. (to be published).

Zysina-Molozhen,L.M., Dergach,A.A., and Kogan,G.A.(1977). Experimental investigation of heat transfer in a radially rotating pipe. HGEEE High Temp. 14, 988.

INDEX*

* Certain expressions appear extensively throughout the text and in
  these instances the index cites key locations only.

parallel-mode rotation, 15,
113
pseudo pressure, 44
Prandtl number, 45
turbulent, 118

radial equilibrium, 34
Rayleigh number, 45
Reynolds number, 40
rotational, 107
Rossby number, 40
rotating, reference frame. 12
rotation, orthogonal mode, 20,
157
parallel-mode, 15, 113

Seider-Tate, 89
straighteners, 133
stress, 24
shear, 25
normal, 25
square tubes, 137
developed flow, 137, 142
laminar flow, 137
turbulent flow, 147

thermosyphon, 83
thickness, boundary layer, 55
total derivative, 25
transition, 170, 179, 201
translating, frame of
reference, 12
turbulent, dissipation, 116
flow, 25, 113
Prandtl number, 118

uniformly heated tube, 113
unit vector, 15
unweighted mean temperature,
52

viscosity, effective, 116
vorticity, 23, 27

water, 119